T. S. Arthur

Nothing but Money

T. S. Arthur

Nothing but Money

ISBN/EAN: 9783744711203

Printed in Europe, USA, Canada, Australia, Japan

Cover: Foto ©berggeist007 / pixelio.de

More available books at **www.hansebooks.com**

NOTHING BUT MONEY

A Novel.

BY
T. S. ARTHUR.

AUTHOR OF "LIGHT ON SHADOWED PATHS," "OUT IN THE WORLD," ETC.

NEW YORK:
CARLETON, PUBLISHER, 413 BROADWAY
M DCCC LXV.

Entered according to Act of Congress, in the year 1865, by
G. W. CARLETON,
In the Clerk's Office of the District Court for the Southern District of New York.

NOTHING BUT MONEY.

CHAPTER I.

"HAT is the price?"

The speaker was a young woman with a small basket on her arm, in which was a steak, a bunch of asparagus, and half-a-dozen eggs. She had lifted a rose geranium from the stand of a market gardener, and stood looking at its pink and white blossoms with admiring eyes.

"Twenty-five cents," replied the man.

The words seemed to make the flower-pot heavier, for the hand that held it went down suddenly, like a scale on receiving additional weight, and the geranium took its place among the verbenas, pansies, calceolarias, phlox and petunias on the flower-stand.

A shade of disappointment fell over the young woman's face.

"Take it for twenty," said the gardener.

There was only a silent shake of the head, as the young woman turned away.

"Fifteen!"

She lifted the flower again, pressed a leaf between her fingers, and inhaled its fragrance. That sweet impression on her sense was the concluding argument. The geranium was hers.

Let us observe this young woman a little more closely, as she moves homeward with a light step, carrying her market-basket in one hand and her flower in the other. Her face is pretty and girlish. Round, blossom-tinted cheeks, fresh and pure; soft blue eyes, full of nestling loves, and bright with hopes that look sweetly confident, onward toward the coming years. Her dress is of pink and white calico, plain in the body and fitting her form closely. The skirt has no trimming, and is short, displaying a small foot and ankle. The foot is covered with a morocco slipper; black ribbons are crossed on the instep and tied around the ankle. A light blue ribbon binds her slender waist, two long ends falling at the side and fluttering in the fresh June air, as she goes tripping on her homeward way. A cottage bonnet of straw, simply trimmed with a band of puffed ribbon, throws her face into half shadow, and gives it a softer beauty. She is twenty — no more.

There is nothing striking or unusual in the appearance of this young person; and yet, as she passes along, one and another turn and glance back for a second observation. There may be two reasons — the harmony and good taste manifest in her plain attire, and the calm sweetness of her fair young countenance. Thus far, life has been to her a pleasant experience. But this need not be told — you see it at a glance.

Not far away from the market-house stands a small

brick dwelling, two stories high. It is new, with a white front door, and green venitian shutters at the windows of both stories. The street door opens into a little parlor. We will take an inventory of the furniture. On the floor is a red and green ingrain carpet, and six drab-colored Windsor chairs look at each other from the opposite walls. A small mahogany table; a narrow mantel glass, flanked by two tall lamps; brass andirons and fender, with shovel and tongs; and a pair of conch shells in the fire-place — complete the attire of this first and best room. Next to it is the breakfast and sitting-room. On the floor of this is a rag carpet, woven in red, yellow, green, and white stripes. A small pine table, stained red, standing in the centre of the room, is covered with a snowy table-cloth and set with breakfast-things for two. Four Windsor chairs, a mahogany work-stand, a pair of broad-bottomed brass candlesticks on the mantel-piece, and a few unimportant trifles, complete the furniture of this apartment. Opening from this room is a small kitchen; above are two chambers, and above these two attics.

The dwelling and its furniture are humble. We have the abode of a young married couple, beginning the world, according to their means, in the most unambitious manner. The rent of this house is one hundred and fifty dollars a year; the entire cost of the furniture three hundred. Enough had been laid up by the young husband to pay for the furnishing, and still have two or three hundred in reserve.

It is early morning, not much beyond six, and Adam Guy sits by the window awaiting his wife's return from

Lexington market. Lydia's small soft hands are to prepare his breakfast, for they have as yet no domestic. As he sits by the window, we will take his portrait. Age about twenty-five; the firm thin lips, slightly falling brow, and cold calculating eyes, plainly indicating the lapse of some years since his departure from light-hearted youth. Our young husband is a man in all that appertains to an earnest life-purpose. He has already measured himself with the world, and girded up his loins for battle.

We cannot say that we like the expression of his face, as he sits by the window waiting for his young wife's return. It is in repose, and expresses some habitual state of mind, or to speak more accurately, the quality of some habitual state. His eyes, half closed, are looking forward upon life — not observant of anything external. There is something hard — we may wrong him to say cruel — in his inflexible mouth. If it was even a little sensual and voluptuous we might like it better. How very cold his face is! Perhaps the dark complexion may have something to do with this appearance. We cannot say; but it is cold and calm. The blood seems never to have found its way there, giving a rich warmth to cheek and brow. All is of the same hue, from forehead to chin. We wonder if it lights up when feeling is at play?

Yes, for it has lighted up suddenly, and all that looked repellant has fled. He has started from the window and is at the door, where his young wife is standing.

"Are you not late, dear?"

He draws out his watch and glances at the time.

Ah! that sudden smile which looked so sweet around his mouth is fading quickly.

" What time is it? "

Do you see the brightness toneing down on her face as her eyes dwell on his countenance.

" Half past six; and you know I must be at business by a little after seven."

His eyes fell upon the geranium.

" Isn't it sweet? "

The young wife plucks a fragrant leaf and hands it to her husband. He does not smell of it, but tears it to pieces in an absent way.

" It is half past six, remember, Lydia.'

He could think only of the business that awaited him.

She goes past him with her face a little paler. She has felt more than we have perceived. The coldness has struck downward with a chill.

In fifteen minutes, during which time Adam Guy has walked the floor across parlor and breakfast-room with unceasing tread, the meal is served. It consists of coffee, bread and butter, and boiled eggs. The geranium is on the work-stand, and its fragrance filled the room.

" Isn't it sweet? " The wife has poured her husband a cup of coffee, and he is busy with his eggs and bread and butter. " And it cost only fifteen cents. The man asked a quarter."

The young husband turned his eyes upon the flowers.

" How much for that thing, did you say? "

" Fifteen cents."

" Umph! What else did you buy?"

"A steak, half-a-dozen eggs, and a bunch of asparagus."

"What did they cost?"

"Let me see — twenty-five, seven, and ten — forty-two cents."

"And the flower fifteen?"

"Yes."

"So your flower cost nearly half as much as your dinner."

Adam Guy shakes his head after a very sober fashion.

"But that's the way of the world, my dear," he adds, in a moralizing strain, and with more of severity than kindness in his tone. "Ornament, beauty, and superfluity are permitted to consume half of almost every man's substance. People would get along easily enough with their necessities. It is the burden of superfluous things that makes so many stooping shoulders. Now, we must be wiser than all this, Lydia. We must not let outside glitter and show bewilder us. There's no use in that flower, which has cost as much as three loaves of bread, or your half-dozen eggs and bunch of asparagus."

Tears are in Lydia's eyes, but her husband does not observe them; her appetite is gone, but he fails to notice the fact. His thought rests in the importance of making her feel that only life's necessities are to be regarded — that money is a thing of too much importance in the world to be wasted on trifles.

"Money is a great power, Lydia," he goes on. "If we have money at command, and to fall back upon, we can be independent of everything and everybody. But, if we waste our money, instead of keeping it for use we will be the slaves of every changing circumstance. Money is a reality, and abides. If you lay it up you have a well-grounded assurance of finding it all safe in the time of need. But your pretty flowers wither up and die, or, living, are a care, and useless. I have made it a rule, for years, not to buy anything that I did not really need. Some men can't keep a dollar to save their lives. They not only spend foolishly all they receive, but their covetous eyes lead them into debt for things that are wholly useless. Such men you find always in trouble. They complain about not being able to get along in the world, and never seem to comprehend the fact that the fault lies at their own door."

"Flowers are not useless things, Adam. God made them."

There is a pleading tone in the low, tender voice of Lydia Guy, as she looks across the table at her husband.

"And he made the rivers also; but that is no reason why we should turn them from their courses, and let them sweep in destruction through our dwellings," is triumphantly answered.

"There is no harm in flowers. They destroy nothing. But, on the contrary, restore to the mind much that is lost in our jarring life-experiences. They are God's messengers of love to our hearts."

"All poetry, Lydia! All poetry! Your flower, there, has destroyed fifteen cents — a thing actually demonstrable. Three loaves of bread would have given us blood and muscle for work in the world. But that poor trifle! Pah! Its useless."

Lydia is neither strong-willed, nor given to contention. She does not, therefore, urge her view of the case, but sighs, and remains silent. Adam Guy talks on, having the argument all to himself, and rises, at length, from the breakfast-table with the air of a man who has settled a favorite point beyond all controversy.

"Throw your geranium out of the window, dear," half laughing, half in earnest, as he kisses his wife at parting. "It has already brought you more trouble than it is worth five times over."

CHAPTER II.

IT was beyond the reach of Adam Guy's imagination to picture his young wife sitting tearful, or in sad, half dreamy abstraction, for an hour after he went away, and all for what he had said about a useless flower. Would his thought have grown tender toward her, if he had known the truth? Would he have chided himself, for letting so small a matter come in to mar the happiness of a young heart, that was beating so true to love and him? No. His thought would have grown sterner, and he would have approved to himself all the coldly-wise sentiments which had been spoken. He would have felt angry toward the flower, which he had only despised as worthless.

Yet, so it was with Lydia. She had a true woman's sensitive appreciation of all things beautiful in nature. From a child, she had been a lover of the earth's bright and beautiful children, the flowers. They spoke to her in a language not understood by grosser natures; and, in their presence, she felt like one lifted into some purer sphere. To hear the flowers contemned by lips, whose words had come so often in music to her ears — from lips to which her ear must bend and listen in all her after-life — ah, that was no light thing! We do not wonder at her tears.

Then, there seemed to her such a hard, cold, calculating vein, in what her husband had said — a spirit not seen before — an intense worldliness — a bowing down to the worship of the lowest and most external things — and an elevation of money as the greatest good. Suddenly, there had come to her a new revelation of his character — not that he had never spoken of economy and prudence — of the repression of vagrant desires, and the folly of waste. These were his favorite themes; and, as he had usually presented them, her thought approved. She would have liked a little more of the ornamental in her household — a few things of beauty for the eye to dwell upon. But, her husband was poor and, in conforming to his circumstances, she felt a sweet pleasure.

Now, a veil had dropped from her eyes, and she saw him in a clearer light. She had thought him earnest and absorbed in business — ambitious to make his way in the world — prudent, calculating, strong-willed, and resolute to do his part in life efficiently. But, she had not understood him as he now appeared in her eyes — a worshiper of mammon, and a despiser of even beauty, if it could not be an offering on this shrine.

When Lydia Guy took up the day's burden of duties again, it felt heavier than before. Her voice did not break forth into snatches of song, as her hands busied themselves with household cares; nor did her feet bear her with their usual springy tread. A shadow had fallen on the distant landscape: she could not see in it the beauty once so delightful to gaze upon; its odors did not steal in waves of sweetness on her sense — it had caught a shade of dreariness. Alas, for the young heart, when brightness fades from its sunny future! The life of Lydia Guy was opening into a new experience.

She had, after arousing herself from the depression of mind occasioned by her husband's fuller revelation of himself, cleared off the breakfast table, and set her house in order. An hour or two for needlework came in at this part of the day, and Lydia was sitting by her little table, when a visitor made her appearance in the person of a young married friend. Warm and tender was the greeting that passed between them; for they were heart-companions.

The visitor's name was Lena Hofland. Her husband was a young physician, who had just opened an office, and like the husband of Lydia had all the world before him, in which to sow and reap. The two young men were standing in life at about equal advantage, so far as worldly goods were concerned. Less than a thousand dollars covered the full amount of Guy's possessions at the time of his marriage; and Hofland's means exceeded this only by a few hundred of dollars. One difference existed, which will be regarded as in favor of Guy; he had earned and saved his seven or eight hundred dollars, while the twelve hundred on which doctor Hofland ventured out into the world, was the remnant of a small legacy on which he had depended while studying his profession. Guy's was an accummulating, Hofland's a diminishing fund. And there was still another difference — Guy had no taste to gratify; no artificial wants;

no expensive weaknesses. A silver dollar, in his eye was more beautiful than a picture, a vase, or a jewel, or any thing desirable only on the score of ornament. He cared nothing for music, and the beauty and sweetness of nature made no strong appeals to his inner consciousness. But Dr. Hofland had an acute perception of the beautiful, and tastes, that, if indulged to their full extent, would have drawn largely upon the amplest fortune. On money he set no value except as a means of procuring that which mind and body required in the world. Applying the metaphysician's distinction — he looked at gold subjectively, rather than objectively.

The young men had been acquainted from boyhood. Their parents were neighbors, and they had attended school together. One from choice, entered a store, and the other, also from choice, became a student of medicine. At this point in their lives there was a divergence in feeling as well as in pursuits. Guy did not lack mind; he had as clear and strong an intellect as Hofland, and in any profession would have been his peer — stood above him, perhaps. His choice of a mercantile life came from no peculiar fondness or fitness for the pursuit, but simply from a money consideration of the case. He saw through trade, the surest road to wealth, and took that road in consequence.

So much, briefly, of the two men and their antecedents. The reader has already comprehended them. They are types of two great classes, to both of which the world owes much. They do not seem of much account in their day of humble obscurity. Their spheres of life are narrow, their places in the world unnoted, their influence scarcely perceived. Strong men, men of gold, men of intellect, if from any cause their eyes happen to fall upon them, hold them in light consideration. They see not an oak's great promise in the acorn's tender shoot, nor dream of the imperial river as they step, without a thought of its limpid waters, over the slender rivulet.

The wives of these young men had been early friends, also. But their tastes were more nearly accordant. Yet Lydia was not so clear-seeing, nor so strong-willed as Lena; else she would never have given her hand to Adam Guy. Lena would have penetrated more deeply into his character; would have comprehended his quality better — and she would have had decision enough to turn from him resolutely, when he approached as a lover.

A flood of light made radiant the face of Lydia Guy as her young friend entered. Their lips met in a heart-warm kiss; their arms went fondly twining about neck and waist. Lena held a small bunch of choice flowers in her hand.

"They are for you, dear," she said, after the kisses had been given, and words of love exchanged.

"Oh! how sweet! Thank you, Lena!" and Lydia received the offering, gazing on it with eyes that felt and drank in the beauty that held them.

"The Doctor bought me a lovely bouquet as he passed through the market this morning, and I have divided it for my friend," said Lena.

"It was so kind in you!"

As Lydia said this, she turned her face partly aside.

A thought came in to mar the pleasure of the moment; to steal away the fragrance that was breathing upon her lips. Almost involuntarily, her eyes wandered to the geranium that stood, yet, upon her work-stand. Lena's husband had made her a gift of flowers; but Adam Guy had blamed her for buying a single blossoming plant, with which to beautify her home.

"They are very sweet," she added, as she commenced examining the flowers that made up the cluster in her hand. "How the odor of this mignionette takes me back along the way of girlhood; and I see the woods and fields again, by the magic of sweet briar and myrtle. What a delicate tint is in this rose! And this bud! Oh! is it not exquisite? White, crimson, soft fading pink, purple, and golden. What a power there is in beauty, Lena! In color and grace of form united. They speak to some inner sense, and that sense responds in thrills of pleasure."

As she ceased, a faint sigh came fluttering through her lips.

"Ah," she continued, "if the useful and the beautiful in life were more closely united. But the hard, stern, plodding useful, persistently separates itself from the beautiful, or spurns and tramples it under foot."

"It is for us to unite them," said Lena. "The doctor and I were talking of that very thing this morning."

"Beauty is costly," remarked Lydia, "and we are poor."

There was a shade of depression in her voice.

"And cheap, also," answered Lena. "A flower is not costly, and yet, in nature, there is no other form of such exquisite grace and delicate proportion; and all the riches of color are added; and all the sweetness of perfume. If taste is genuine — if our love of beauty is, indeed, a passion of the soul — then may we find perpetual enjoyments, even though our lot in life be poor and humble. A true lover of art may enjoy a statue or a picture far more than the owner. Speaking on this very subject, the doctor remarked a day or two ago, that the love of possessing works of art, was inferior to the love of art, and that therefore, the man of true taste, though unblessed by fortune, might enter into higher pleasures than those to whom wealth brought every desire of the eye. If I look at the picture in a rich man's gallery, and carry it away in my thought, am I not its owner in the highest sense? Fire cannot destroy it; misfortune cannot bear it away; no accident can mar its fair proportions. It hangs in the gallery of my soul, among other precious things, and the outside world has no power over it. It is mine, though his ownership cease; mine, though the paint and canvas are borne away to the antipodes."

"You are growing into a philosopher," said Lydia, smiling.

"Yes, thanks to my good husband. He is helping me to get up higher, so to speak; to breathe in a purer mental region; to see things in their best relations. We are poor, you know, but the doctor says, that we may be as happy in our poverty, as the rich in their riches; nay happier, for we are free from the temptations of the rich. The lesson we have to learn, is that which teaches a moderation of desire. Wants must be few, and not too often told. We must cultivate a love for the beautiful, rather than a love of possession; and learn to see beauty with an interior vision."

Ah, how different all this from the uttered sentiment of Lydia's husband! He looked only to possession — to wealth as the best gift the world had to bestow. Beauty, in comparison to gold, was nothing. He spurned it as unlovely! The contrast, now so strongly presented, almost saddened the heart of Lydia.

"You are not so bright as usual," said her friend.

Lydia smiled, and tried to look happy. But the light did not linger sweetly radiant in her countenance. It faded out slowly.

"How is the doctor succeeding?" asked Lydia, changing the subject.

"As well as might be expected," was the reply. "He has been called to one or two good families, and if he should be liked, their influence will be of great use to him."

"Will his income be sufficient for your expenses?" inquired Lydia.

"O dear, no! So far, his paying practice has not been at the rate of three hundred dollars a year."

"You say 'paying practice;' has he any other?"

"Yes, and a full share of it among the poor."

"Ah! Is that so? Does he attend the poor for nothing?"

"There is sickness among the extremely poor, as well as among others; and the physician cannot refuse to visit and help the sick because they have no money to recompense his services. We happen to have many very poor people in our neighborhood, and the doctor is called in frequently. It is a Christian duty to attend them, and one from which he cannot hold back. They are God's patients, he says, and he is so largely a debtor to God that he must take all opportunities for payment."

"They ought to recompense him in something, if it were ever so small," said Lydia.

"How are you to live? The laborer is always worthy of his hire."

"The Doctor has still five hundred dollars on which to draw. This will carry us through a year; beyond that we trust in a good Providence."

"Not a very encouraging prospect."

"We push aside discouraging thoughts," was replied. "To-day is ours, and we try to get all the happiness from to-day that it has in store. This the Doctor calls life's true philosophy. I get a little nervous, sometimes; and look into the future in a spirit of doubt. What is the result? Doubt peoples the future with forms of evil, and my heart grows faint as I look at them. But, when I turn back to the present, I find myself surrounded with blessings; and I lift my heart in thankfulness. 'Only to-day is ours, Lena,' my husband will say, when I question about the time to come, 'only to-day is ours, to work in and enjoy. Let us do our work faithfully, and take the enjoyments, and God will see that our to-morrows are all right.' It does me good to hear him say this. He is such a consistent, right-hearted man, Lydia! His thought is so clear, that I see it always, when expressed, as the utterance of truth. It does not come into my heart to question what he says. But I am only talking of what interests me. How is Mr. Guy — and what are your prospects in the world? How is life looking in the far away future, to which the eyes will turn with asking glances?"

" My husband is not so easy in mind as yours," replied Lydia, " Though his present condition and future prospects look more promising. His salary has been raised to twelve hundred dollars, and it will not cost us over six or seven hundred to live at the outside. Then he confidently expects to receive an interest in the business of the firm where he is employed. Give him that position, he says, and he will consider his fortune made — will, to use his favorite expression, ' snap his fingers in the world's face.' "

" I am glad to know that everything has such an encouraging aspect," answered Lena, with genuine pleasure. " You ought to be very happy."

Lydia sighed faintly, as her eyes dropped to the ground. That fair promise in the future, did not fill her desires. There were intuitions in her soul, that pictured something more, yet left a trembling fear of disappointment. This day was to be memorable in the history of her inner life, as one on which her mind had awakened to a new consciousness touching her husband's character, and its want of harmony with her own. What Lena had just said of her husband, was as a foil bringing out to clearer perception the opposite characteristics of Adam Guy — and they were unlovely in her eyes. Not that he had all at once revealed himself in his true aspect. Lydia had failed to read the signs aright. They puzzled her at times; and she often questioned as to a meaning beyond anything dreamed of in her estimate of the man to whom she had committed all things that were holiest and most sacred. But to-day the veil dropped from her eyes. That brief scene with the flowers was a revelation; and she stood no longer a questioner, or in doubt.

CHAPTER III.

YDIA did not feel more peaceful for this morning visit from her friend. Some things that Lena said, particularly about her husband, remained distinctly in her thoughts. The promise of this world was fairer for Adam and herself, than it was for Doctor Hofland and his wife; but the promise for happiness was on the other side.

At dinner time, as Adam Guy and his wife sat at the table, on which their meal was laid, Lydia referred to the call she had received from her friend Lena.

"Playing the lady," said Adam, sententiously.

"How? What do you mean?" Lydia did not clearly understand her husband.

"Aping rich and fashionable people," replied Adam, "in morning calls, when she ought to be at home attending to her house, and aiding her husband. She keeps a servant, which will cost for hire, board and waste, not less than two hundred dollars in the year — more, I'll warrant you, than the Doctor's practice will yield him in that time. Now, I don't call any woman, who is so idle and extravagant, a good wife. Instead of helping her husband to succeed, she will help to keep his nose always on the grindstone. That is not like you, dear."

This last approving sentence, spoken in a gentler tone than he had used in condemning Lena, softened the shock of language that was felt as a harsh and unjust judgment.

"It is not as I am doing, Adam," she returned, "But, all are not alike in this world, you know."

"And all do not come out alike. As we sow in this world so will we reap, Lydia. Can thrift come of idleness, waste, and extravagance? Never! The Doctor and his wife are beginning wrong, and they will come out wrong — mark my word for it! Lena is just as able to do the light work of their household, as he is to meet the demands of his profession. Does he hire a man to make his pills and spread his plasters? I have no patience with women who commit such folly! Gadding about the street, and making morning calls! Pah! It nauseates me, this pretence of gentility. I thought better of Lena. Why, if you were to set up to play the lady after this fashion, Lydia, the house would soon be too hot to hold us. I wonder at the Doctor for submitting to such a state of things."

"It is all right in his eyes, I presume," said Lydia.

"Then they're a couple of blind fools. That's the best I can say for them."

As the young man remarked this, his eyes lighted on the bouquet of flowers which Lena had brought for her friend, and which had been placed in a glass of water on the mantelpiece.

"Ha! More flowers! Where did they come from?"

"Lena gave them to me." The blood crimsoned over Lydia's face.

"Umph! Bought 'em, no doubt."

"The Doctor bought them for her, as he went through the market this morning."

"And she gave them away. Upon my word! she valued her husband's gift."

"She only divided it with me," replied Lydia. "I love flowers, and she wished to give me pleasure. It was kind and thoughtful in her."

Adam Guy shook his head, in marked disapproval.

"And so it was the Doctor who threw his money away? Well, they are a precious pair! I wonder where they expect to come out?"

"Right in the end," said Lydia.

"They will, when arithmetical laws change, and subtraction gives the result of addition — not before. But, the world is full of such people. Just look at it, for a moment. The Doctor has only three or four hundred dollars to come and go upon, outside of the returns from his practice. At the best, his practice will not give him over five hundred a year, on an average, for the next three years. Very well; look at it, as I say. Look at it. House rent, two hundred; cost of a servant, two hundred more; table expenses, three hundred, at the lowest figure; clothing and outside expenses, two hundred and fifty; flowers, jewelry, pictures, gewgaws and other nonsense, two hundred more; in all, eleven hundred and fifty dollars! Run this through three years, and you have three thousand four hundred and fifty dollars. Now, let us see what the prospect is for meeting so large a sum. There is, we will say, four hundred to start with; and we will give six hundred a year for the Doctor's average income during the next three years, and that is a liberal estimate. Three times six hundred make eighteen hundred — add four hundred, and we have two thousand two hundred dollars of means against an expenditure of three thousand four hundred and fifty! Figures don't lie, my dear. At the end of three years the Doctor will be twelve hundred and fifty dollars in debt! Think of that!"

A troubled expression came into the fair young face of Lydia Guy, as she sat looking at her husband. She understood him perfectly, and saw, for the time, clearly with his arithmetical eyes. The promise was not a good one for her young friend. There was misfortune in the world for Lena, and the heart of Lydia was touched by it, in saddening anticipation.

"Debt! — yes debt, that curse of a man's life!" resumed Guy, almost bitterly, as if he felt the fiend's grip on his arm. "They will be overridden by debt, as surely as the breath is in them! Somebody's money besides their own will have to go for their waste and extravagance. Whose shall it be? Not mine, I can tell them. No, not a dollar of mine! Adam Guy's hard earnings and careful savings shall never go to sustain the pride, self-indulgence, and wasteful extravagance of such people. I'll burden nobody, and nobody shall burden me. I have the industry, patience, self-denial, and persistence needed for accumulation, and with it the nerve to keep what I gain. No man shall find in me a weak spirit of yielding. I can be iron and brass to importunity;—and I will."

There was a tone in her husband's voice, and an expression on his face, that made the blood flow back in a chill from Lydia's heart. She had never seen him in just the light he now presented himself.

"You haven't thrown that flower out of the window," he said, with more than half seriousness, a little while afterward, as they arose from the table, and his eyes glanced toward the geranium which his wife had bought in the morning.

"No, nor have I any intention of doing so," replied Lydia. "That would be wantonly to destroy a thing of beauty."

"There's no use in it," said Guy.

"I'm not so sure of that. The sight of a flower refreshes my mind. If I am dull, a new life flows through my veins; if I am sad, a cheerful spirit awakes. Don't condemn the flowers, Adam; they have a mission for our hearts."

"And that mission is, to teach us how frail and perishing is all ornament — how valueless are flaunting color and mere exterior grace! We spend our substance for naught, when we spend it for these. That is the lesson the flowers teach us, Lydia, if they teach us anything."

He took up the flower-pot, as he closed the last sentence, and poising it in his hand said:

"Let me throw it from the window."

But Lydia sprung to his side, and catching his arm, cried —

"No! — no! — Don't do that!" in such earnest remonstrance, that he desisted from his purpose. She felt that her husband was going too far, and anger blended with the feelings that made her heart beat more violently, and sent the hot blood to her face. Ere the flush of anger died, she said, with a sharpness that stung him:

"Adam! You are stepping a little beyond your prerogative. If I care to have a flower, it is not for you to object."

"It is for me to object to a foolish waste of money," he answered in a cold, firm voice; "and I advertise you here, that I shall always do so."

And saying this, Adam Guy took up his hat, and left the house.

The day which had opened so unfavorably for their peace, gathered blackness as it advanced. Here was the first storm that had troubled their serene sky. Lydia stood, for some minutes, like one who had been stunned by a blow. Then she sat down — not in tears, but with a pale, abstracted face, and brows knit gloomily. Painfully the conviction forced itself upon her mind, that there had been a great error in her girlish estimate of Adam Guy's character; that she had comprehended him only in part. The morning's troubled questionings were taking the shape of distinct perceptions. She saw him as she never had seen him before, and felt herself removed, as it were, to a distance from him. A sense of repulsion arose in her heart. The moral beauty, which had appeared as a fair garment clothing his spirit, seemed to fade and change to an unlovely vesture. If it was with him, as from this new revelation of himself, it appeared to be, the sweet idea she had formed of a marriage union, would prove to her like the airy fabric of a vision. Their minds could never grow into each other, by the attraction of similar tastes, feelings, affections and principles — could never blend into harmonious oneness. All this and more, was seen and felt by Lydia, as she sat lost to external things for a long, long time, after the departure of her husband.

Lydia Guy was alone in the world, so far as near relatives were considered. Two years before her marriage, the death of her mother had left her without a home, or any means of support beyond the product of her own hands. From school, she passed to the work-room of a dress-maker, and in six months learned the art of constructing garments so skillfully, that she was able to support herself in independence. Not alone did her fair countenance, grace of form, sweetness of manner, and more than ordinary intelligence, attract the eyes and win the heart of Adam Guy. These would have allured him in vain, had there not appeared the more solid basis of thrift and industry. He saw that she would make a good wife, in another sense than is always considered; that she would work and save, and help him to grow rich. He did not find in her the nonsense, frivolity and want of thought that displayed itself in so many of the young ladies who came under his observation; and he was especially pleased to note the fact that she had acquired a better estimate of money than is ordinarily held by her sex. The necessity of earning before spending, had produced this result.

Before marriage, they had talked freely about their housekeeping arrangements. Lydia noted, that in his calculation of expenses, nothing was said about the hire, or cost of keeping a servant in the beginning. As she felt well and strong, and really desired to join hands with her future husband, as a "helpmeet for him," she saw no objection to this; she was willing, in the outset, to perform all the work of their little household. It could be a labor of love, and nothing else. She was used to being busy over some kind of work all the day long; and the thought of having a home of her own to work in, and one loved above all others, to work for and make happy, was imagining to herself a paradise.

And so they had begun their housekeeping, as we have seen, Lydia doing all her own work; and, up to the day on which she is introduced to the reader, doing it cheerfully. But, from that day, "a change came o'er the spirit of her dreams."

CHAPTER IV.

NEARLY all the afternoon, on that first day of Lydia Guy's introduction to the reader, did she sit with idle hands, dreamy eyes, and lips just touched with a shade of sadness. The stream of her life, which had, since her marriage, been dancing along musically in the sunshine, all at once left the open fields and gentle declivities, losing itself in sluggish pools that widened and diverged, and hid their dark depths under thick, shadowing trees, and tangled brake. She could look forward in sweet hope no longer. There was cloud, and obscuring night, on all the future, that a little while ago had been so full of promise. The life, into which her consciousness was opening, had a strange, repellant aspect; and a shudder crept into her heart as she tried to see light and beauty ahead, but could make out nothing, distinctly, among the gloomy shadows that obstructed her vision.

Must all beauty, all gentle charity, all the soul's loving worship of things that dwell in regions above mere sordidness, or hard, accumulative actuality, be crushed out? No wonder that a shudder crept through her heart! No wonder she sat with idle hands through that afternoon, trying to comprehend all that was meant by this new relation in which she found herself standing to life. Was money — property — material wealth — the greatest good? Did it comprehend all worth living for? Must everything else be cast down for its enthronement — beauty — friendship — charity — love — all the heart's riches? Was she not to have a flower, even, because it was no money-producing, or money-saving instrument? or because it absorbed a few pennies or dimes? or, worse still, in the eyes of her husband, fostered taste, and a love of the mere beautiful, which were expensive attributes.

The heart of Lydia Guy rebelled against all this. If such were her husband's requirements, she did not see how it was possible for them to draw any nearer in spirit — to grow into that sweet oneness of life which her maiden fancies had loved to dwell upon as including the highest of human felicities. Her talk with Lena in the morning in no way tended to reconcile her to this change in the programme of life. How sharply in contrast stood the character of her husband with that of Doctor Hofland? What a light seemed to hover over the home of her friend, while clouds were gathering in the sky that arched above her.

A few warm words had passed between Lydia and her husband at dinner-time, and he had gone away with a stern, admonitory sentence on his lips. He had spoken with authority, and left a spirit of rebellion in the heart of his wife. The law of force had come in, setting aside the law of love and sweet compliance was at an end. Foolish man! Blind, weak, besotted man! For what dross was he bartering the rich red gold of life!

Not with that light step, which gave to every motion a grace, as on the afternoon of the day before, when Lydia made preparation for her husband's coming at twilight, did she move, now, as the shadows began to lengthen, in the work of providing their evening repast. A heavy heart makes the feet heavy.

Adam Guy was not a man from whose feelings any ripple passes quickly. All disturbances went down deeply, and surged to and fro for a long time after the cause had ceased. It must be remarked, however, that he was susceptible of disturbance only in the direction of his avaricious cupidities. Lay your hand on these, and he felt the jar long and profoundly. Assault these in never so small a degree, and sympathy, pity, tenderness toward the assailant, even humanity, died out instantly. He was armed and guarded at every point.

And so, Adam Guy's feelings did not soften toward his young wife during the few hours that elapsed from the time of his parting with her at mid-day, until he met her again in the first fall of twilight shadows. He saw in her a weakness that must be crushed out. His hand was upon it, and come what would, he meant to extinguish its life. Germs of extravagance were beginning to show themselves, which must be robbed of vitality. Not a single word of the sentence — "Adam, you are stepping a little beyond your prerogative! If I care to have a flower, it is not for you to object!" — failed from his memory. He conned them over and over, and over again, each time rejecting them with a stern purpose.

"Stepping beyond my prerogative!" — so he talked with himself, ever and anon — "We shall see! That was unwisely said, Lydia. Not for me to object to waste and extravagance! Indeed! I wonder who is to object, if not I? Please heaven, I will object to the last; and not only object, but extinguish waste and extravagance. If this comes from the introduction of a single worthless flower, I shall take good care that my house is not transformed into a conservatory. Forewarned — fore-armed."

Wearied with beating about in a vague uncertainty; weak and bewildered; the heart of Lydia began to lift itself toward her husband, as the day declined, with a yearning for the sunshine of love which clouds had hidden. She repented of her hastily spoken words, and even went so far as to remove the geranium, which had been the exciting cause of this trouble, from their sitting-room. In what spirit would he come home? That was the question of greatest concern now. Would he bring the hard, threatening, almost angry face that frowned upon her in parting; or the old, pleasant face, in which she read so many tender meanings? Oh, she could not live without love! could not go on through life in a spirit of antagonism. No! — no! She was not strong enough for this. Death were to be preferred.

And so, looking away from the causes which had wrought this unhappy alienation, she tried to let reawakening love for her husband cover the hard, bare, unsightly aspects of character which had suddenly revealed themselves; and in this spirit she was awaiting his return, when she heard his well-known footsteps crossing their little parlor floor. She was in the kitchen, busied with preparations for supper, but came forth quickly, meeting him as he entered the sitting-room, where the table was spread. The light in her eyes, and the smile on her lips, died **instantly**; for the face of her husband was cold as a sea of ice. The usual kiss was intermitted. Adam did not offer it, and the heart of Lydia rebelled against solicitation. A few distant words were exchanged, and then Lydia went back to the kitchen, with a heart of lead in her bosom.

Almost silently passed the evening meal. Adam sat through it with a cold, implacable face — Lydia with a heart ready, at each moment, to gush through her eyes.

After the meal was finished Adam took a newspaper from his pocket and commenced reading, while his wife removed the tea things. As she went and came, passing from their sitting room to the kitchen, he glanced, furtively, over the edge of his newspaper at her face, and was a little surprised as well as annoyed, at seeing tears falling silently down her cheeks. It was the first time Adam had seen his wife in tears, and it made him feel rather strangely. This was something not taken into the account in marriage. He had bargained for smiles, not tears; for a mind that should be in complete harmony with his own — never in opposition; for a second self. What was the cause of these tears? That question came naturally, and Adam Guy answered it from his own stand-point, and selfish perceptions.

"And so it has come to this!" he said, speaking to

himself. "Because I will not consent to waste and extravagance I must be punished with tears. But it wont do. Adam Guy isn't the man to be turned aside from what is right by a woman's tears. If Lydia wont hear to reason — wont act like a prudent, sensible wife — the trouble be on her own head. As to wasting my hard earned money on such nonsense as flowers, it never shall be; and the matter may as well be settled first as last. As for Dr. Hofland's wife, I don't think her the right stamp of a woman for Lydia, and I'll break off the intimacy, if possible. Hofland is an extravagant, thriftless fellow, and his wife is just like him. He is out of my books, anyhow. I don't like the way in which he is beginning life — living beyond his means, and making debt certain. In less than a year he'll be on the borrowing line. There'll be a break between us then, just as surely as the sun shines, should terms of intimacy exist. The prudent man forseeth evil, and I am a prudent man. He is going his way and I am going mine, and the roads have a wide divergence."

Thus he talked with himself, fortifying his mind against his wife and strengthening his purpose to have his own will in all that concerned them.

"What's the matter?" he asked, in an abrupt, half imperative, half surprised tone, as Lydia came in from the kitchen, after having completed her duties there, affecting just to have made the discovery that she had been weeping.

The heart of Lydia was too full; she could not answer either calmly or indifferently, and so made no reply. On this silence Adam placed his own construction, and placed it wrongly of course. It was from moody self-will, that she did not answer — so he read the reason.

"A woman doesn't cry for nothing. What's the matter? What's gone wrong? Are you sick?"

Lydia had drawn a chair up to the little work-stand, on which a lamp burned, and near which her husband sat with his newspaper in his hand. She had already taken some needle-work into her lap. The tone in which he asked these questions, only made a reply on her part the more impossible; and so she bent her head down over the sewing she had taken up, and gave no response. This, to Adam, was like placing an obstruction in a flowing stream. The waters of his mind became agitated, and bore onward in turbulence.

"Can't you speak? Haven't you a tongue?"

Lydia started, looked up at her husband in a bewildered way, and then burst into tears, sobbing violently.

Adam Guy was at fault. He was dealing with an unknown element. A woman's mind is a mystery to most men — most of all to men like Adam Guy; yet have such men, in their blind antagonisms the fatal power of wounding to the heart's core. He sat, coldly observing the agitation of his wife, until her sobs gave way to an occasional short spasm in breathing, and these at length to low, fluttering sighs.

"I think, Lydia," said Adam, in a firm voice, when the storm of feeling had completely died away, "that you have permitted a very small matter to come in, and make itself a disturber of our peace. I objected, as I claim that I have the right to do, to waste of money in any form. I objected to the purchase of useless flowers; and I still object. You charged me with stepping beyond my prerogative. That was unwisely remarked, permit me to say, and does not set well upon my mind. You threatened to do as you pleased, and I warned you against such folly, and again warn you. I am not a man to turn aside when I know myself to be in the right; and I am very certain of my position in this matter. I am a poor young man, with my way to make in the world. I earn my money by hard and patient industry, and cannot see it thrown away in trifles. You know my opinion of flowers. I gave it to-day; and it is, and will remain, unchanged. Money spent in them I hold to be worse than wasted. A bouquet fades in a day, and the money it costs might as well have been cast into the sea. Pot flowers are a constant care, and involve waste of time, in addition to waste of money — and time is money. So, you see that I have common sense and prudence on my side, opposed to weak fancy and extravagance. I'm sorry you have taken so small a matter into such serious account — that you have made yourself unhappy about a poor geranium. Now, let me beg of you to rise above all girlish weaknesses of this kind, and be a sensible woman — all in earnest as to life's true objects. There are more enduring things than flowers to be gained, Lydia. Let us see eye to eye — let us keep step in our onward march to a high place in the world — let us deny all mere self-indulgences, that are unsatisfying at best, and always enervating, and press forward to the attainment of real and abiding things. Let us spare now, to spend in the coming time, when we can afford to spend."

Adam Guy paused. His young wife was bending closer over her work, than when he commenced speaking, and her hand moved steadily and quickly. She did not look up, nor answer.

"Am I not right? Does not your own better judgment approve what I say?"

But she made no response.

"Lydia!"

She looked up, showing a pale face and red eyes.

"Why don't you answer?"

Her eyes, with an expression in them so strange, that he scarcely knew them as the eyes of his wife, looked steadily into his. But there was no reply on her lips.

"Have you nothing to say?" demanded Guy.

"Nothing." How calm and cold her voice! It gave not a sign of feeling. Her eyes fell away from the eyes of her husband, and went back to the sewing in her lap. The needle hand, which had paused with the thread half drawn, took on again its quick and steady motion; and there was silence between them through all the evening hours — silence and alien thoughts.

CHAPTER V.

NOT far away from the humble home of Adam Guy, was the more tasteful dwelling of Dr. Edward Hofland. It stood a little back from the street, with a garden and shrubbery in front, arranged in the neatest order. The house was called a half-house, standing with the end to the street, the front door opening directly into the parlor — a side door led into a narrow entry, from which the stairs ascended. On one side of this entry, was the parlor; on the other, a small dining or sitting room, and next to this the kitchen. The house was but two stories high, with an attic, and all the rooms were small.

For this modest abode, the doctor, paid two hundred dollars a year. Its garden in front had pleased his taste, and he had taken it in preference to one with quite as much room, which would only have cost one hundred and fifty dollars. This fact was known to his old friend, Adam Guy, who had blamed him as extravagant, in no choice terms — not to himself, of course, but in conversations with his wife. A reason, besides the gratification of his taste, and one which did not come into the thought of Adam Guy, had also influenced the doctor in making his selection. As a professional man, success depended, in some degree, on social appearances; and he was very well satisfied that the more tasteful looking house would prove the cheapest — and he was right so far.

Let us look in upon the doctor and his wife, on the evening which passed so drearily, and with such a bad promise for Adam Guy and Lydia.

Doctor Hofland was a quiet, thoughtful, rather grave looking young man, just a little above the medium height, slender, of fair complexion, and clearly cut features. His eyes were brown rather than blue, and dark for his complexion. They were grave, like his face, but, like his face, kindled beautifully when thought grew active, or his feelings warmed. His whole air was refined — his manner quiet, — gentlemanly.

The young wife of Dr. Hofland was of a different temperament from her husband, and as different as to personal appearance. She possessed a clear, strong, resolute mind, which was under the discipline of sound, good sense, ardent but not blindly impulsive feelings, a cheerful disposition, and a warmly affectionate nature. She had a round full face, complexion dark, eyes black, full of light, and all alive when she spoke. You would not call her features regular, but would say —" How beautiful!" at the first glance. Lena was a charming young woman, the favorite of all who knew her, and the tenderly loved wife of an appreciative husband.

We may always know something of people's characters, by the things with which they surround themselves.

Swedenborg says, that, in heaven, the scenery and objects by which angels are surrounded, and even the garments they wear, are representative of their states, and change always as their states change. The same thing is true of men and women in this world, only the changes here do not take place immediately, but by gradual progressions — mind acting but feebly on the hard substances of nature, and moulding them to its ideal shapes by slow degrees. In a man's dress — in his house and furniture — in all material things, that he selects and arranges as the ultimate forms of his thought and affection, we see images of himself, and comprehend therefrom the quality of his mind. As his character changes, we see corresponding changes in his exterior things.

In the furnishing of Adam Guy's house, the man's character was clearly illustrated. Love of money was his ruling desire, and to this everything had to be subordinated. Mere ornament, in his eyes, was a superfluity — a useless waste — and so only the plainest and least costly articles were bought. There was nothing, out toward which taste could go, and rest in tranquil delight.

Let us see how it is with Doctor Hofland. An ingrain carpet is on the floor of his small parlor, but it cost ten cents a yard more than that of Adam Guy, the extra ten cents having been paid in consideration of a finer quality and more tasteful pattern. The chairs, instead of being solid "Windsor," are "cane-seat," black, with ornamental gilding. Instead of a cheap mantle glass there is a small French clock; and in place of the two glass lamps, a pair of neat china vases, that rarely miss their burden of fragrant flowers. A pair of card tables stand on opposite sides of the room; and in front of the fire-place, with its shining fender and andirons, is a small center-table, covered with books. Three or four choice engravings ornament the walls. How clearly does every article, as well as the style of arrangement, indicate the mental quality of the directing mind or minds. Here, husband and wife acted in sweet harmony, and their home gives expression to their mental states. In the other case, Guy's will governed in the selection of nearly every article, and in his home you saw an outbirth of his state and character only. In the things by which he had surrounded himself and wife, Lydia's mind did not rest in calm content. Taste and feeling went out in a restless search for objects in fuller correspondence with themselves, and came back weary and dissatisfied. A few seemingly unimportant changes in the style of their furniture — a trifling, and not costly addition here and there, a little more of the "uselessly ornamental," and it would have made all the difference in the world to her. But Adam Guy saw in all this but weakness and folly. The useful only had attractions for his eyes; and what he meant by the useful referred to the mere wants of natural life, as the basis of effort toward worldly accumulation. Whatever came not in this category was superfluous, and to be rejected. He could subsist on husks, but not so the more delicately organized nature of his wife. On the fare that would sustain him she would feel the tooth of a perpetual hunger, and her life, only half-developed, beat about in restless, weary, unsatisfied consciousness of defect — halting, astray, and stumbling in weakness and sadness to the very end.

A lamp was burning cheerfully on the centre table in Doctor Hofland's little parlor, and his wife sat by it sewing, when the doctor came in from a professional call. Tea had been waiting for some time.

"You are late, dear," said Lena, as she received her young husband's fond kiss on almost bridal lips.

"Yes; I went to see a poor woman on Fell's Point. I am attending a Mrs. Helme in Exeter Street, and she told me about her mother down on the Point, who was very ill, and begged me to go and see her; and I could not, of course, refuse. God's poor are always with us, and we cannot turn from them, when they stretch their hands toward us, and be conscience-clear — at least as a physician, I cannot."

The tea-bell rang at this moment, and they went to the dining-room, where, their single domestic having supplied everything for the table, they were alone.

"I didn't tell you that I called to see Lydia this morning," said Mrs. Hofland, as she handed her husband a cup of tea.

"No; how is she?"

"Very well."

"And happy as the day is long," remarked the doctor.

"I don't know about that," replied Lena, a slight change in the sunny glow of her face. "I can't think it possible for a woman of Lydia's peculiar character to be happy with a man like Adam Guy."

"He was never a favorite of yours."

"O dear, no! His sphere has always chilled me. My freedom is gone in his presence. I feel like a blossom shrinking in a frost-breath."

"But, Lydia found in him a congenial spirit. All are not alike."

"I cannot but feel," said Lena, "that, in wooing Lydia, Guy assumed a character not in agreement with his true quality; and to-day I thought I could detect signs of an awakening to a hard realization of the fact that their lives were not, and never could be, in harmony."

"Adam loves money," said the doctor.

"And means to accumulate it. Lydia said that his salary had been advanced to twelve hundred dollars."

"Ah! I'm pleased to know that. The doctor spoke with genuine pleasure.

"And it will not cost them six hundred to live, Lydia says."

"So they will be six hundred dollars better off every year. A comfortable prospect."

"And, moreover, Adam anticipates an interest in the firm. Give him that position, said Lydia, and to use his own language, 'he will snap his fingers in the world's face.' Now, doesn't that express the man's true character in a sentence? Snap his fingers in the world's face! He's selfish to the core, Edward — selfish to the core! And, as the sure consequence, unhappy. I told Lydia how hopeful and cheerful you were in your small beginnings and clouded future; and she said, with a sigh, that her husband was not so easy in mind. How can he be? Love of money, and the narrow spirit of selfishness, which always accompanies it, are never satisfied with the present, nor resigned as to the future. Men like Adam Guy are always impatient in the present, because gain comes too slowly for their desires, and restless about the future, lest their one great cast in life should fail."

"He will snap his fingers in the world's face!" said the doctor, in a tone that mingled surprise, regret, and disappointment — as if an unpleasant revelation had come to his mind. "I don't like that, Adam Guy. Snap your fingers in the world's face! No man can afford to do this. No man is true to himself; far less true to society, who lives in that spirit. I knew he was selfish, and a money lover, but I hardly thought him so blind and foolish in his selfishness as this. Alas for him, and alas for his wife, if his words express a true purpose! Life will prove to him, and it may be to her, also the saddest of failures. The *life* is more than *meat* and the body than raiment. What is simple possession — what is wealth — if made *more* than the life? A burden and a curse — nothing less, nothing more, as thousands, if the heart spoke out, would testify. And yet, the thousands who succeed them go on in the same blind, besotted way — stifling the soul's higher instincts, dwarfing its powers, suppressing its yearnings after the things for which it hungers and thirsts with an immortal desire — and for what? Just for gold — for gold, and the unsatisfying good that gold can buy — this, and no more — no more. If Adam is going to walk in the broad way to misery — to misery in this world, I mean, for there is no happiness by the way nor at the end — I pity him from my heart."

"But most of all, I pity Lydia," said Lena. "If Adam will make his bed of husks, and put thorns in his own pillow, let him enjoy them if he can; but for Lydia! Ah, my heart grows faint for her. She is of a purer substance, and of a tenderer organization. She will have no sordid loves to sustain her—no end of worldly gain or worldly pride, like him; and so must endure or die. This marriage is a serious thing, Edward. Thank God, that you are not like Adam Guy! Could we be happier in a palace than in this modest home of ours? Would hundreds of thousands of dollars make our hearts beat in closer unison, and fill them with a purer happiness?"

"Not unless our lives were in accord with all things good, and true, and beautiful. Not unless in our souls were the spiritual riches to which this outward wealth corresponded. Anything less than this, and the exchange would be to our loss, instead of gain."

"So I feel, and say, thank God, that you are wiser than most men!" And Lena threw tender glances upon her husband.

"It is one thing to see clearly the right way in which to walk, and another thing to go forward in that way," said the doctor. "I can philosophize, but do not find myself living up to the philosophy I approve."

"That is the experience of every one," answered Lena. "Our ideals must always embrace unattained good, or there would be no going on toward perfection. But, in our contrasts with others we are able to see the positions we occupy. Take Adam, for instance, proposing to snap his fingers in the world's face so soon as he is rich enough to care for nobody, and contrast your ends of life with his, as expressed in that declaration. How much higher you stand! You are wiser and better than that, my husband—wiser and better than that, thank God!"

CHAPTER VI.

AFTER tea, Doctor Hofland spent the evening with his wife, reading and conversing in their little parlor. Patients were not in abundance yet, and he had time on his hands. They talked of many things, and dwelt, with hope and interest, on their future. Like Adam Guy, Doctor Hofland had visions of advancement in the world; of success in his profession; of accumulation. He looked forward to the day when a widely extended practice would give him a liberal income, influence, and position—looked forward, selfishly, as all men in whom natural life has not become subordinated to a spiritual and regenerate life, look. But, unlike his friend, Adam Guy, his thought did not centre upon and revolve only around himself. He had generous thoughts and purposes toward others—humanitary ends—aspirations that included the common good. Sordid love of money was not an element of his nature. He had no desire to accumulate, merely for the sake of riches, and the selfish independence of the world their possession would give. As thought went forward to the time when he should have money at command, and influence among men, he loved to dwell on embryo schemes of social good — benevolent, educational, or industrial. Means to ends, he did not see clearly. That time was yet to arrive. He was young and immature. But the germs of good citizenship were in his heart, and fructifying life was beginning to stir their latent forces with a prophecy of things to come.

"If I were only rich!" How often did this sentence fall from his lips, as he looked on poverty and suffering, or contemplated the mental and moral destitution around him. And there were times when, in the ardor of his desire to relieve want, or help forward in some good enterprise, he fancied himself free from selfishness, and willing to devote all his powers to the service of others. In this, though it was but an ideal state of good, there was given a reward. Into even the desire to benefit others flows a blessing — how much higher the blessing for those who make desire an ultimate actuality.

"If I were only rich!" There is not a moment of time in which this aspiration does not rise from some heart dissatisfied with the amount of possession God has given. "If I were only rich!" said Doctor Hofland, as he sat in his little earthly paradise that evening, "the world should be better for at least one life. I would not hoard my money for spendthrift heirs, nor mortuary endowments — but scatter blessings as I passed along. Rich men are God's almoners. Alas! how few are conscious of their responsibility, or dream that a day of reckoning must come."

The doctor's mind was excited, and his imagination fast bearing him away. But, a word from his wife drew him back again, and his wings drooped from their airy flight.

"God only requires a use of the talents given," said she. "Are we not all almoners in our sphere of life?"

"Truly said, Lena! and I stand reproved."

"No, no, not reproved." There was a tone of deprecation on Lena's lips.

"Corrected, then, darling. Thank you for clipping the wings of my too aspiring imagination. It is even as you have said; God only requires a use of the talents bestowed. I am rich! Rich in the power to do good. I have but to dispense, freely, according to the ability He has given. Like to many others, I look away from my present sphere of life, and long for a wider field and higher opportunities. But, if not faithful in what is least, how can I expect to be trusted in greater things."

"Ah, if we could always keep that thought in mind, how much more of peaceful life would be ours. Faithful to-day. Let that be our motto, Edward. Faithful to-day."

The eyes of Dr. Hofland turned from the face of his wife, and a sigh fluttered softly on his lips.

"Is not that the right doctrine?" Mrs. Hofland leaned toward her husband, and laid a hand gently on his arm.

"Yes, darling. It is the true doctrine. Faithful to-day; and an impressive sense of its truth has sobered me. Faithful to-day! Ah! it is this looking beyond to-day — this living in our to-morrows, that is such a hindrance to useful life. Our powers do not come down with that will into the present, which is needed to give them true efficiency. We reserve strength for the future, instead of putting it all forth in our to-days. Faithful to-day. You have expressed life's true philosophy in its simplest formula. Let us accept the axiom as our rule of conduct. If our present work is always taken up and faithfully done, we need have no anxiety about the future. As servants of the Heavenly Master, whose hands never lie idle, the right work for us to do will be given in the right time. He knows what is best for us, and best for those to whom good is to come through our life in the world."

"I do not think," said Lena, "that we shall ever be happier than now. Oh, is not life sweet to us!" And her bright face grew sunnier. "God was good to me when he put love in your heart, Edward. I pray to become worthy of your love."

"If eye sees to eye, and heart beats to heart, darling, ever as now, life shall be to us one long, sweet day of happiness," returned the young husband, breathing the words on Lena's lips. "There will be care and toil; hope and disappointment; sorrow and pain — but, with a love in our hearts growing purer, stronger, and more heavenly in its origin all the while, we shall never sit in darkness — shall never be comfortless."

"Purer and more heavenly," said Lena, as her eyes expressed deeper meanings, "the words bring back what our minister said yesterday, that true marriage was a union of souls. 'What God joins together,' he said, had a significance deeper than came to the common apprehension. God conjoins in marriage by means of spiritual affinities, and these are heavenly. Without a good life, he argued, no true interior marriage was possible. There might be a likeness and a nearness of souls from natural affinities; but genuine interior marriage — that conjunction which made of two minds, male and female, one harmoniously pulsating individuality, for all time and eternity — only took place with those who, through obedience to God's spiritual laws, advanced from natural into spiritual life along the gradually ascending way of regeneration. I have thought about that new doctrine of marriage, Edward, a great deal. Shall we be thus conjoined, dear husband? That would indeed be heaven!"

"And I have thought of it, also," replied Dr. Hofland. "Our minister was right in what he said. The truth of his words came like sun-rays, bearing illustration into my mind. God must join together by spiritual affinities, and these are heavenly affinities. May He do his good work in our hearts, Lena, that we may be one forever!"

CHAPTER VII.

THE heart never loses its memory. Every experience records itself so indelibly, that, always, what has preceded in our lives throws its shadows or sunshine on succeeding states. We cannot forget, if we would.

Lydia could not forget. Alas, no! The record of that day, when she awoke, suddenly, to the truth in regard to her husband's character, was an uneffaceable record, engraven, as with a pen of iron; and, in all her after life, from the sad, disheartening page, not a line or word was obliterated. And other records followed — sadder and more painful — followed in quick succession, as Adam Guy put off concealments, and let his true quality and ends of life appear without disguise. He was a hard, resolute man, and trampled on all weaknesses as obstructions in his way. In his first conflict with Lydia he saw that she possessed certain traits of character that might be difficult to manage, and had a certain reactive force, the ground of which was not understood. But it was no part of his system to study a case, and with shrewd diplomacy, adapt himself thereto, gaining his ends by Jesuitical craft. He moved forward in directer lines, bearing down opposition by the force of an imperious will. He knew only the bend or break system. And so, the conflict begun, there was on his side no furling of banner, nor sheathing of sword. It was war to the bitter end.

During the first year of Adam and Lydia's married life, reactions on her part were, from the out-reaching necessities of her nature, frequent; but always she had to retire, with a heart bruised, bleeding, and palsied, from the contest. If she gained in anything, it was at a cost so far beyond the gain, that conquest was a defeat instead of a victory. Day by day, and week by week, Guy became more and more absorbed in money-making. Sooner than he had expected, an interest in their business had been tendered by his employers, and he was throwing himself into the vortex of trade with an abandonment of thought and purpose that dwarfed all other considerations. In the beginning they visited old friends, and had evening visitors in return. A concert, a lecture, or some public entertainment, was, now and then, added as recreation, though the cost of these made Adam rather indifferent to them as sources of pleasure. As thought hovered, more steadily, in circles around his leading end of life, he grew more and more indifferent toward all things not ministrant to his avarice, and, before six months had passed away, rarely stepped beyond his own threshold, after coming home from the day's employment.

"I'm too tired to go out," was the stereotype reason offered, when Lydia suggested the return of some friendly visit. Indifference was the true reason. Sometimes he would utter expressions of dislike toward the persons mentioned. In fact, this dislike of people was a feeling that gained on him steadily. Lack of thrift in a man condemned him utterly. No matter what other qualities he possessed — no matter how gifted or useful in the exercise of his talents, or kind of heart, or self-denying for the good of others — if he lacked the quality of thrift, our young merchant despised him.

It can readily be seen how, with such feelings as Adam Guy possessed, he would naturally separate himself from all close friendships. These might be entangling! Few men were as self-dependent, and as earnestly given up to the work of money-making as he was, and, therefore, nearly all around him were in danger of stumbling by the way; and he did not care to have relations with any one of a character to warrant applications for a helping hand in emergency.

Doctor Hofland, his old friend, called in now and then, with his wife, to sit an hour or two in the evening. While the meetings between Lydia and Lena were tender and cordial — seasons of real heart-enjoyment — Adam held the doctor more and more at a distance, and rarely made any response when the latter spoke of his profession and prospects. His own business Adam never intruded, and if the conversation led him to make any reference thereto, it was of a vague and rather discouraging character. He did not wish to have the doctor know that he was beginning to accumulate, lest he should want some help.

"We must go around and see Doctor Hofland and his wife," said Lydia one evening, several months after the date on which our story opened. "It's more than eight weeks since we were there, and they've called twice during the time."

"I don't feel like going out," was Guy's answer.

"We mustn't let feeling always influence us," replied his wife. "Come! I want to see Lena to night. And I know you will enjoy an hour with the doctor. He's always so bright and cheerful."

"It's more than I know, then," replied Adam. "The fact is, I don't fancy the doctor half so much as I once did. He's getting prosy."

"Prosy? Why, Adam! I don't know a more interesting talker among all our acquaintances."

"He doesn't interest me. Very little that he converses about comes within the range of my interest. I grow dumb in his presence."

"Why, Adam! How can you say that?"

"I do say it, and it's the truth. Doctor Hofland or I have changed very materially in the last year or two. The fact is, he's getting too wise — in his own conceit, I mean. He likes to show off what he knows — to talk largely; and that doesn't set well with me. I hate pretension, and always did."

Lydia felt a choking sensation in her throat. She made no reply, and her husband went on.

"And besides, I don't like the way he's living. It isn't honest."

"Why, Adam! How can you speak so of the doctor?" Feeling sent a glow to Lydia's face.

"Because I believe just what I say. No man is honest, in the right meaning of the word honesty, who lives beyond his means, as he is doing. He's going behindhand every day, and knows it; and yet, denies himself nothing. Every time we've been there he's had some elegant new books to show, or some costly engraving, or some silly trifle of a parlor ornament. Only yesterday I met him with a package in his hand, and he told me that he'd just been buying a choice English edition of some new work which I had never heard of. And then, I don't believe a week passes over their heads that they are not at some place of amusement. All this costs money, and somebody will have to pay the piper. It wont be me, though, I can tell them! Adam Guy knows better than that how to take care of his money. Let them go their ways, and we will go ours, Lydia. Doctor Hofland is pretty nearly off of my books, and so is his wife. They're well mated, and will pull evenly to the edge of some precipice, and dash over together. Well, let them, if they fancy such episodes in life; it'll not trouble me. I've got my own way to make, and shant bother myself with the insane conduct of other people."

Lydia sighed, and was silent. Every act and sentence of her husband had come to be the turning of a leaf, on which she read a phase of his character; and in all phases, the one likeness of a sordid regard for money was never absent. She saw it in every word, and sentence, and act. If she opposed this love of money, she was hurt in the contact, always. It was the ruling purpose, that set aside every opposing thing, and ignored on bare suspicion.

They did not go to Doctor Hofland's on that evening,

as Lydia had desired. Feebly she rallied to the argument in defence of Lena, but was borne down by an intemperate dogmatism that dealt in excessive condemnation.

The truth was, apart from Guy's growing alienation of feeling toward Doctor Hofland, resulting from causes already apparent, he did not approve the influence of a character like that of Lena upon his wife. Lena, in his eyes, was a worldly, extravagant woman, with ideas wholly averse to true home enjoyment; and her influence over Lydia, who was strongly attached to her, could, in his eyes, only prove injurious. Already he had noticed a change in Lydia's state of mind whenever she received a visit from her friend, or spoke of having called to see her; and the change was adverse to contentment.

So little of a yielding spirit had Adam Guy shown, after the honey-moon—so little of deference to her tastes, feelings, or wishes, where they impinged in the least upon his darling love of money—that Lydia had learned to be wilful and persistent in some things—to require concessions that love, or her husband's thoughtful consideration, would not have made. Thus, in three or four months after their marriage, she spoke of hiring a single domestic. There was no response on the part of Adam—he heard, but did not answer. His manner, however, was not to be mistaken. Lydia understood him fully. He could not approve. And yet, for all this, the domestic was employed, and Adam had to submit. He took his revenge, however, after the manner of such men, by wearing a clouded brow, and putting on a chilling reserve toward his wife, that robbed her days of sunshine, and made the nights dewy with tears.

So in other matters of minor concern, appertaining to Lydia's domestic life, dress, friendships, and expenditures, she learned, by painful necessity, to act in open or concealed opposition to her husband in many things; yet, almost always, the opposition cost her only a little less suffering than submission. There were necessities of her nature, free impulses, tastes, that could not be wholly denied. Life would have gone out if some aliment had not been furnished to these.

CHAPTER VIII.

THUS, as we have seen, Adam Guy, as fortune began to smile on him, commenced the work of separating himself from the world as to any personal feeling or interest. The world was a foreign nation, with whom it was not safe to have any entangling alliances. Commercially, and solely for his own advantage, he was ready to hold relations with this world, and did hold relations — but here, the laws of trade and his own shrewdness protected him. Friendship and personal interest were outside to all these, and of an entangling nature; and so he cut them off.

"If I go in a boat with a man that cannot swim, and a storm capsizes us, I may drown in attempting to save him." So Adam Guy reasoned the matter with himself. "It is safest, therefore, to go alone, or not go at all. If I attempt to pass through a wilderness country with weak and sickly men, and they give out by the way, I cannot leave them to perish without the world's execration; so I will go, full armed and provisioned, alone. Let every man take care of himself. That is my doctrine. If every man makes himself safe and prosperous, the world will be all right, and its affairs go on bravely. The idle, the vicious, the extravagant, and the wasteful, mar all the harmony of things, and pull down faster than the most earnest workers can build. A fortune that it takes a life to accumulate, is wasted, often, in a year. But, no man shall spend and waste for me. Adam Guy will see well to that."

It will thus be seen, that Guy's separation of himself from old intimacies and associations was not the result of a passing idea, nor based on a growing indifference — but the effect of a settled principle of action. Money was more to him than friendship, and so, friendship was thrown overboard.

"A babe in the house is a well-spring of pleasure;" so says the proverb; and rarely is the sentiment falsified. The springs of pleasure were running low in the house of Adam Guy, when a babe came with the sweet airs of heaven, odor laden, around it — came in sunshine to a clouded home — came to kindle love-fires that were burning feebly.

Adam was naturally fond of children, and this fondness, when it became stimulated by parental affection, warmed into a tender solicitude that filled his heart with a new delight. This babe he could love safely; so he felt, even if the thought did not take a distinct form in his mind. He could mould the young being to his will, and make it as sordid, and money-loving, and accumulative as himself, and so a safe companion for the time to come.

In another year another babe came, and another followed at the close of a third annual cycle. At the close of ten years six children made music and discord both in the home of Adam Guy — discord for the most part. Four were boys and two girls.

During this period of time Adam prospered in worldly matters. The firm in which he was a partner had largely extended its operations, and made heavy profits. Through the retirement of an inefficient member of the house, he had been advanced to a higher and more controlling position, with an increased interest in the business.

Ten years had wrought many changes in Adam Guy. You would hardly have identified the self-important, yet complacent gentleman, who quietly negotiated with you to-day for the purchase or sale of half a cargo of sugar or coffee, as the same individual who, in the position of junior clerk, opened the counting-room door ten years gone by, and bowed, respectfully, as you passed in to transact your business with his superior. He has developed rapidly during these ten years — but all in one direction. He is a keen, eager merchant, and nothing else worth speaking of. A money-making machine, with all the new improvements attached. The whole force of his life has gone in one direction, and made him strong, shrewd, and far-seeing in all things appertaining to his ruling desire. But, this development has been at the expense of all other faculties and endowments. He knows business, and the things appertaining thereto; but outside of this he is ignorant and feeble-minded. Touch him on the common intellectual topics of the day, and you will meet with no clear response; get his opinion on a question of domestic government, and its lack of common sense will surprise you; on the education of children — on home management — on right conduct in life — on man's social duties — on taste, art, or literature — and the man's utter want of perception, judgment, and information, will stand out in singular relief. And yet, he will talk dogmatically, and have his will where rule is possible, though every step be taken with crushing force, and hearts bleed under his iron heel. Where the love of money comes in and rules the man, he becomes an implacable tyrant.

Another change we must note. The strong desire for money which filled the heart of Adam Guy, extinguished, as we have seen, all pride of appearance in the beginning. But, mercantile pride, as wealth began to accumulate, stimulated personal pride. Adam Guy, the merchant, was beginning to stand out before the people, and Adam Guy must look to his style of living. The little house, with street door opening into the ten by sixteen feet parlor, might do for Adam Guy, the clerk, but it was hardly the thing for Adam Guy, the merchant. Something was due to appearances. So, two years after his marriage, a larger house was taken — one with a passage running through the front building, and separating two parlors with folding doors. Three hundred dollars were spent in additional furniture. But, in making this addition, avarice was in steady conflict with taste and pride. Avarice contended for cheapness, and avarice conquered in almost every instance, though Lydia was always on the side of taste and pride, and got wounds and bruises not a few in the contests.

After the pain in parting with money in taking on a more expensive style of living, had in a measure subsided, our young merchant found a certain poor compensation in contrasting himself with others. On his way to and from his business, he passed, daily, the modest little home of Doctor Hofland, with whom he maintained the outward signs of friendship when they happened to meet — their visiting courtesies had ceased long ago — and something of contempt for its meanness, mingled with his pride. His prophecy in regard to the doctor's certain embarrassment had come true. Income and expenditure had not been rightly adjusted, and debt was the consequence. Once, under extremity, the doctor ventured to ask the tempoary loan of a hundred dollars. He did not get the money, and never made a second application in that quarter.

For the space of five years, Adam maintained himself with but few additions of furniture, in this second home. Lydia, whose maternal duties were blended with household cares, conducted her increased establishment with notable economy, yet never to the satisfaction of her husband. He supplied her but meagerly with money, and forced her, in consequence, to make bills at grocers, dry goods dealers, shoe makers, and so on, and scolded roundly when he had to pay them. He grumbled if fuel needed replenishing, and insisted that wood and coal were burned unnecessarily. A hundred mean little things were done in the line of economies that we cannot stop to particularize, and which made him contemptible in the eyes of servants, and, we might almost say, in the eyes of his wife also, from whose heart genuine love had long ago departed. Adam Guy was a dictator and a supervisor in his household—not a loving thought-taker for the comfort, contentment, and happiness of its inmates. He ministered to the wants of his family with a grudging, and not a liberal hand, and seemed to feel that every gush of free laughter among the children, or sign of pleasure on the face of his wife, was, in some incomprehensible way, a draught upon his guarded coffers — and must be answered by a frown.

The intense selfishness of Guy reproduced itself, by a natural law, in his children. All children are born with selfish inclinations; but, according to the ruling desires and mental habits of their parents, which are reproduced in offspring, these inclinations are modified and counter-balanced in ways innumerable. Where a father gives himself to a single sordid idea, as was the case with Adam Guy, and pursues it steadily, day by day, carrying not only disregard, but contempt and dislike of others in his heart, he will transmit to his children similar inclinations, which, if not weakened by opposite things from the mother, or repressed by discipline and education, must render them selfish in the extreme, and, as a consequence, antagonistic to each other. There cannot be love among such children. Their world is home, and they will, by the force of an inherited determination of soul, incline to have interests separate from this world, and to guard these interests with jealous care; nay, as selfishness is at heart a robber, and they have no salutary fears of law, or social well-being to restrain them, trespass and wrong must follow; and it did follow among the children of Adam Guy. They were in conflict with each other from the beginning.

Their father had no judgment in regard to family rule, and comprehended only the law of force. Of love, as a power, he had no conception. His discipline, therefore, was oftener hurtful than salutary. They did not seem to love, but to fear him; and, as a consequence, always felt in a state of opposition to his commands. Thus the temptation to disobey was always before them, and they felt the power of few restraining influences.

Adam, the oldest, inherited his father's love of accumulation and hoarding. He was individualized among the children, in this particular, from the earliest period in which character began to manifest itself. If he received a cent or a fi'penny bit, he dropped it into his money-box, instead of spending it for candies or a toy. In this, he showed a quality after his father's own heart, and received more paternal commendations for his saving propensities, than for anything else. Thus love of money was stimulated, instead of being wisely repressed, as it always should be, where it shows itself in a child. The right use of money — the sparing to spend wisely — the love of accumulating for some useful end — these should always be taught and cherished; but the miserly spirit never! For that will curse the possessor through life, and may destroy his soul eternally.

John, the second child, had less of his father's decided characteristics. He resembled his mother in person, and mentally, showed many similar features; but, selfishness was as predominant as in his older brother, manifesting itself not in avaricious hoarding, but in spending all that came into his hands for his own gratification. He never shared with his brothers or sisters. A cake, a candy or an apple, would be eaten and enjoyed by him in sight of their longing eyes, and not a crumb or a slice be divided with them! In this, he was unlike his mother. It was an image of his father's selfish soul, the transmitted activities of which were only modified by new elements of character derived from her. In him was shadowed forth the selfish spendthrift.

Lydia, the third child, was bright, active, and self-willed. A resolute trespasser on the rights of all, and almost always in sharp conflict with the rest, she had no true sensibility, no conscience — using the word, as we often apply it to children who manifest little or no moral sense — who are not truthful nor honest — no native kindness of heart — no love of anything out of herself. Edwin, Frances, and Philip, the three youngest children showed leading characteristics quite as distinctly individual as their oldest brothers and sister. These will appear, as their lives are developed in the progress of our story.

The mother's influence over her children was not of a decided character. It was irregular, indeterminate, and impulsive. She had lost her way in life and pressed forward, by a kind of necessity, through desolate and unfamiliar regions, with a never removed pressure of concern upon her breast, and a never appeased hunger in her heart. The masculine strength of character on which she was to lean — the wise intelligence that was to be her guide and polar-star — these she had not found. Adam had proved, instead of the tender, loving husband, she had thought to gain, a hard master, to whose service she was irrevocably bound. A man the light and warmth of intellectual sunshine. He would have been thought to her love, and thus she would have gained through him a higher region and a clearer vision. But, there was no intellectual or moral wisdom in Adam Guy, to which her soul could adjoin itself — nothing that she could love and rest in, with the confiding truth of her nature; and so, in companionship with him, she was astray, and in the dark, pressing onwards by the force of necessity, yet groping about, eagerly and impatiently at times, and again moving on in pulseless abandonment to what seemed a dark and cruel fate.

No wonder that Lydia did not prove a wise mother, efficient for the training of her children; no wonder that she was weak, fretful, irresolute, and without system in her management. Poor health came to increase her inefficiency. The exhaustion of her system, through the rapidly succeeding births of so many children, added to maternal cares and household duties, so enfeebled mind and body, that she was in no sense competent to fill the place she occupied; and yet, no one could hold it but her.

As in the beginning, Adam's will was the general law of the household, and in his prosperity he continued the same careful supervision of expenditure, treating his wife as if she had no right to a voice in anything where his darling gold was concerned.

Thus it was with them, after ten years of married life. Alas for the golden hopes, which had made all these desolate cycles beautiful in the future of Lydia, on the day she laid her maiden hand so trustingly in that of Adam Guy, and called him husband!

CHAPTER IX.

THERE had been a second change in the external of Guy's life in the world, up to the period of which we are now writing. He accumulated steadily, and pride as a merchant demanded that he should live in better style. He felt himself becoming of more consequence every day — it was purse-proud consequence, the meanest kind of basis on which to build self-estimation — and it was needful, therefore, to assume an exterior, in some degree suited to his mercantile status. Rising men, whom he now met in the walks of trade, talked of houses, furniture, crrriages, and country-seats, in a way that dwarfed his modest house into meanness.

A change followed. From the retired part of the city, in which he had lived for seven years, he removed to a more fashionable quarter, and took a house at seven hundred dollars a year, expending over two thousand in refurnishing. It went something against the grain, this outlay of money and increased expenses, but pride was the goad that pricked him onward.

Let us take some three years subsequent to this change, a closer view of Adam Guy's home. Let us open the door of his fine residence, and go in and sit down with him, amid his wife and children.

It is evening. Through the whole day, Mr. Guy's thoughts had flowed in the one direction of business; and so eager had been the purpose which made these thoughts active, that, in more than a single instance, they struck with disturbing force against hindrances or impossibilities. This was no unusual thing, for the purposes and thoughts of our grasping merchant were always in advance of the orderly results of business.

Adam Guy came home in a dissatisfied state of mind, consequent on several causes. There had been a decline in the sugar market; two or three large cargoes having arrived at New York, prices had receded a quarter of a cent. Their firm held a thousand barrels, in anticipation of a rise. Of course he was disturbed. The difference of a quarter of a cent a pound on a thousand barrels was a serious matter; but, what if there should be a further and heavier decline! Another cause of disturbance was the failure of a merchant in Virginia, who was indebted to them over three thousand dollars. No intelligence had been received as to the character of this failure; but the worst is usually feared in all such cases. Mr. Guy feared the worst. Then a good customer, to whom they had been selling for years, had gone over to a rival house, which had offered eight, instead of the usual six months' credit. But we will enumerate no farther. To a man whose ruling passion was the love of money, and who thought and worked only to that end, these were enough to make bitter for that day the wine of life. And so Adam Guy came home at its close, with knit brows, shut lips, and a feeling of angry impatience in his heart, toward everything that came in his way.

A pale, unhappy looking woman, sitting amidst a group of noisy children, lifted her eyes, half timidly, half hopefully, — as if dreaming of a smile that once came with the music of familiar steps,—to the face of Adam Guy as he entered. His knit brows and tightly shut lips, threw their shadow instantly over her countenance. The feeble light, which had flickered there for a moment, went out. There was no cry of joy among the children as their father entered, but a sudden suppression of voices. He did not speak, but moved to a large easy chair, and sitting down, dropped his chin upon his breast, and let his thoughts go back to his gold and his merchandise. In a little while the children's hushed tones came out again, and filled the ears of their father with a disturbing clamor.

"Silence!" The word came in deep, commanding utterance.

Stillness reigned for several moments. Then low whisperings began, increasing to a murmur, that soon rose to wild discord again. Loudest among the mingling voices, were those of Adam, the oldest boy, and Lydia, the oldest of the two girls. Adam was a favorite with his father, because, of all the children, he showed hopeful qualities. Thrift was foreshadowed in his regard for money. Toward Lydia, on the contrary, he always seemed to bear ill-will. Nothing that the child could do, appeared to meet his approbation. Scarcely an evening passed, that she was not ordered to leave his presence, and, unless she conducted herself with signal circumspection, the same thing occurred at almost every meal. This discrimination against Lydia was regarded by Mrs. Guy as unjust, and she often had sharp words with her husband in consequence, and not unfrequently in presence of the children. On this occasion, as the voice of Adam and Lydia rose, in contention, the father said, peremptorily —

" Lydia ! Go out of the room ! "

" And Adam, do you go also !" spoke out the mother.

Adam looked toward his father, and hesitated ; Lydia moved back a few paces, and then stood still, looking at her mother.

" Did you hear me ? " The heavy jar of Mr. Guy's foot gave emphasis to his word. Lydia started, and receded towards the door.

" Adam ! Didn't I tell you to leave the room ? "

Mrs. Guy spoke sharply. The boy not moving, still looked at his father, and seeing no command in his eyes, remained firm.

" Go, sir, this instant ! " The stamp of Mrs. Guy's small foot was added to her voice.

" Why do you send him out ? " Mr. Guy turned, frowningly, upon his wife.

" Because he's equally in fault with Lydia. "

" No, I'm not ! I didn't do anything ! It's all her ! She's the worst girl that ever lived ! " Thus Adam, reading his father's eyes, and having the memory of past things before him, came out in his own defence.

" She's a great deal better than you are, sir — a great deal better ! " Poor Mrs. Guy's self-control and prudence were all gone. In her weakness, standing alone as she did, she was borne down by the pressure of indignant feelings. Lydia still remained near the door, awaiting the result of this diversion made in her favor. Encouraged by her mother's defence, she flung back upon Adam a stinging retort, which he returned, with interest added. A movement by her father, that Lydia well understood, caused a hasty retreat from this field of unequal combat; she passed through the door, shutting it after her, and retired to a region of greater safety.

"It isn't just to make her the scape-goat of every wrong done in the house," said Mrs. Guy, speaking indignantly, and looking with angry eyes upon her husband. He made no answer beyond a contemptuous curl of the lip, and letting his chin droop again, looked away from the present disquietude to the more important matters of his merchandise. Adam did not leave the room, and was soon engaged in wrangling with his brother John, who, in turn, followed his sister into temporary banishment.

The supper scene was one of usual discord. The undisciplined children were restless, noisy, and contentious, and their father ill-natured beyond his wont. Lydia, John, and little three-year-old Frances, were sent from the table and to bed. Adam, who deserved banishment quite as much as the rest, maintained his place, though vigorously assailed by his mother, and ordered by her to follow his brother and sister. Only his father's word, when his father was present made law for him.

Later in the evening, after all the children were in bed, Mrs. Guy broke in upon her husband's silent meditations on the subject of loss and gain, with the sentence,

"Have you twenty dollars in your pocket-book?"

"No. What do you want with twenty dollars?" Mr. Guy started from his reverie, and moved his person uneasily.

"I have use for it," was coldly answered.

"Didn't I give you ten dollars yesterday?" demanded Mr. Guy.

"Yes; but I paid the milk-man's bill."

"How much was his bill?"

"Six dollars."

"Well, you had four left?"

"I bought sundry little matters."

"Sundry little matters! O yes! Sundries cost more than any thing else. Sundries eat out the life of all prosperity. I would discharge a clerk who made an entry of sundries in one of my account books. Sundries! I hate the word!"

"Well, Mr. Guy," Lydia drew herself up somewhat haughtily, and spoke firmly and with covert sarcasm in her voice. "I will particularize. There were four yards of gingham for aprons — seventy cents; tape, sewing cotton, needles and pins — fifty cents; a pair of scissors — forty-five cents; four yards of bonnet ribbon — a dollar and a half —"

"There, there! That will do!" broke in Mr. Guy, impatiently. "I thought there'd be ribbons, or some sort of finery, in the case. It's money, money, all the while — a regular drain. A man might as well pour water into a sieve."

Mrs. Guy looked down at the sewing on which she was engaged, and made no answer. Mr. Guy kept on.

"I received Yardly & Co.'s bill to-day."

"Well." Mrs. Guy did not look up.

"How much do you suppose it is?"

"I've not the least idea."

"You haven't? Upon my word! If you haven't any idea, I wonder who should? Didn't you buy the goods?"

"I presume so."

"Two hundred and twenty-two dollars and sixty-five cents, madam! A bill, as long as my arm!"

And Mr. Guy drew forth the bill, and displayed it before the eyes of his wife, saying, as he did so —

"Look at that!"

Mrs. Guy, after taking it from his hands, went over, item by item, slowly and thoughtfully.

"It's correct as to the articles and prices," she said, in a quiet tone.

"It is!"

"Yes, sir."

"Then I call you an extravagant woman, Lydia! A bill for dry goods of nearly two hundred and fifty dollars in less than six months."

"There are six children and myself to clothe, Mr. Guy; and if you will glance at the bill you will see twenty-six dollars charged for linen that was made up into shirts for yourself. Now, it strikes me, sir, as being a very moderate account."

"Moderate!" And Mr. Guy, who had taken the bill from his wife's hand, tossed it from him in angry contempt. "It's nothing but money, money, money — morning, noon, and night! I can't turn, but the word money is flung into my ears! I dread coming home, half of my time, for so sure as I cross my own threshold, the cry of money is heard. The horse leech's daughter were a companion to be ——"

"Adam, take care!" Mrs. Guy turned on her husband suddenly. She had been spurred into reaction so many times that a defiant spirit had crept into her heart. Love went out long and long ago.

The tone and look of his wife caused Mr. Guy to pause, and hold back the words that were on the lip of utterance.

"Take care of what?" he growled, ill-naturedly.

"You might go too far," said his wife, with cold resolution in her voice.

"You talk in riddles; I don't understand you; speak out in plain language when you address me." Mr. Guy's tones were contemptuous. But his wife uttered no further word. She had him at bay, and that was sufficient. To contend was no part of her nature, though smarting assault often roused her into temporary reaction, and there was, at times, a quality in her reaction which had so threatening a look to Mr. Guy that he held himself back from a final encounter. It was so in the present case.

Turning himself partly away from his wife, and dropping his chin again upon his bosom, Mr. Guy went back to his gold and his merchandise. A silence of over ten minutes followed, when a servant opened the door of the room in which they were sitting. Mrs. Guy looked up, and seeing who it was, said —

"Very well, Jane; I'll see you in a moment."

The girl, who was dressed to go out, retired.

"Adam, I want ten dollars for Jane," said Mrs. Guy.

"Ten dollars! You don't owe the girl so much as that?"

"Yes; and I promised her that she should have it to-night."

Mr. Guy drove his hand into his pocket, and taking out a purse, selected therefrom a ten dollar bill.

"There!" he said, thrusting it toward his wife. "And I wish you'd have some regard to my demands. I've said, a hundred times, that I wanted the girls paid every week. Don't let this occur again."

"Give me a certain reasonable amount regularly, and I'll see that everything is paid as I go along," returned Mrs. Guy.

"What do you call a reasonable amount?"

"A sum equal to our household expenses."

"That's very indefinite," said Mr. Guy.

"You know about what it costs us in the year."

"I know that it costs us a great deal more than it should."

"Perhaps it does; but that's neither here nor there. Take our expenses for the last year, and divide the sum by fifty-two. This will give you the amount of our weekly expenses. Place that in my hands regularly, and I'll see that you are not annoyed by these constant demands for money."

Mr. Guy did not respond. The proposition had often been made before, but he had a fancy that his wife would, under this arrangement, spend with a more liberal hand, and run up heavy dry goods' bills into the bargain. He loved his money too well to trust it with her in any sums beyond tens or twenties; too well to let it pass from his hands without grave intimations of its value. With the instinct of his avaricious nature, he saw that if he set apart a certain sum weekly, and handed it to his wife, as a thing of course, she would hold on to it with less tenacity than if every renewal of her purse were attended by remonstrances, interrogatories, and lectures on waste and extravagance. "Women don't know the value of money," was one of his favorite self-justifications; and he acted up to this settlement in all his dealings with his wife.

"I want ten dollars more," said Mrs. Guy, seeing that her husband made no response to her proposition. Her voice was firm, and just a little sharp with impatience.

Mr. Guy dashed his hand into his pocket again.

"There!" An angry frown darkened his face as he handed Lydia another bank bill. "It's nothing but money!" he muttered, almost savagely, as he arose to his feet, and commenced stalking about the room. His wife retired silently.

The scene we have presented is a simple illustration of the home-life of Mr. and Mrs. Guy. Love, as we have said, had died out long ago, and in its place was hard antagonism. Truly had Guy expressed the thought and interest by which they were bounded, when he said — "It's nothing but money!" Beyond or above this the wings of his spirit had not power to lift him; and fettered by his life — bound to a sordid earth-clod — Lydia could not get above the limitations of a sphere to which a base marriage had doomed her. He was ever holding back his yellow dross; she, from the necessities of her position, ever grasping after it. And so, the finer qualities of her nature — all her tastes — all her aspirings after higher things — all her loves and humanities — were stifled, overlaid, or extinguished. The promise of her young, sweet life, was rendered fruitless. Beauty had turned to ashes.

So we find it with Adam Guy and his wife after the lapse of ten years. How few of those who envied their wealth and style of living, dreamed of all the hollow mockery by which they were surrounded. They had money, but at what a price!

CHAPTER X.

TEN years have laid their burdens and their lessons on the hearts of Doctor Hofland and his wife, as well. They have not found it all meadow-path and sunshine. Rough places have wounded their feet, and storms have found them far distant from sheltering rock or hiding covert. But, in all trial, disappointment, anxiety, and affliction their hearts have drawn nearer and nearer together, gaining, at each approach, more unity of pulsation.

A defect in the character of Doctor Hofland was lack of worldly prudence. The absence of all sordid qualities left him in danger of setting too light a value on money; and the absorption of thought in his profession, kept his mind away from a due consideration of life's economies, without which it is almost impossible for a man, solely dependent on his own efforts, to keep himself free from embarrassment. He was too apt to let want, and not means, rule in the matter of expenditure. The tastes of such a man are costly things, and, if gratified, absorb money rapidly.

We find the doctor and his wife living in a pleasant house on Charles street, among merchants, bankers, and men of property, and in a style indicative of professional thrift. His practice has largely increased, — for, in addition to his acknowledged skill, he has personal qualities that render him popular. Pass we into his home. Let us contrast the parlors with those of Adam Guy, who is worth at least forty thousand dollars — the doctor's account with the world, we are sorry to say, stands seriously against him. In the merchant's parlors, we find lace and damask curtains, Brussels carpet, rich mantel glasses, pier-tables, and maroon velvet furniture. These, looking cold, stiff, and stately, suggest only a money value. You think of what they cost — not of their use in the household. No pictures smile down upon you from the walls; no urn or vase, no bronze or Parian statuette, gives grace, tenderness, or human beauty. You might as well be in a cabinet maker's showroom, for all the sentiment of home that is inspired.

You stand now in two smaller rooms, communicating by folding doors, — and a home-feeling is penetrating your heart. These are the parlors of our friend, the doctor. Instead of damask and lace curtains, we find simple Venitian, window-blinds. There are no glasses in the piers, and only a small one on the mantel in the front parlor; a hair cloth sofa stands in one room, and a piano in the other; the carpets are ingrain, and the chairs cane-seat. But, there are many things beyond these, and your eyes go to them instinctively. Here hangs a landscape, that gives you a dream of summer never to be forgotten, and there a home scene of exquisite tenderness. You smile now at the humor of a picture that hits off a foible of character, and take the lesson to yourself — or, more likely, apply it to another. From wall to wall you pass, lingering before painting and engraving, and drinking in beauty and sentiment. Then you turn to examine a small bronze figure of Canova's Dancing Girl, which stands on the mantel, and go from this to an exquisite Hebe. On the two centre-tables, you find rare books, rich in art-treasure — the wealth of European galleries.

But these are not obtainable by any small outlay of money. A piece of canvas, two feet square, may cost more than a gilded mirror; and the wall adornments of two small parlors, like those of Doctor Hofland, absorb a larger sum than all the damask, rose-wood, and French plate in the drawing-rooms of a merchant-prince. Art is expensive. It was actually so in the present case. These pictures, books, statuettes, and other articles of taste, cost more than all the handsome furniture in Adam Guy's parlors. You look grave at this and will look graver still, when we tell you that our pleasant friends, Dr. Hofland and his wife, are enjoying these rare and costly things at the price of debt! Trying to enjoy them, it were better said, — for, with such a drawback, minds like theirs can have no real enjoyment.

Do not blame them too severely, — for, at heart, they are not dishonest. Of set purpose, they would wrong none. Good deeds and kindness have marked each step of their way through life. These costly things which you see, have been gathered, one at a time, all through the ten years that have elapsed since you first looked in upon them. The sum of their price has never been thought of; if you were to ask their aggregate money-value, they would not answer correctly, within hundreds of dollars. Debt has come by gradual advances, year after year, as expenditure steadily exceeded income; and now, when the close calculating merchant is worth forty thousand dollars, the physician is " worse than nothing," by at least three thousand.

As we gave you a near view of the home-life of Adam Guy, after the lapse of ten years, we will now let you see how it is with Doctor Hofland and his wife. They have had five children. Of these, two have passed through the gate of death into heaven. Their oldest, a daughter, named Lena, from her mother, — the second, a boy, — and their youngest, a baby-girl, ten months old, are with them still.

We take the same evening, on which we opened for you the door of Adam Guy's dwelling, and will let you pass into the Doctor's home.

There sits Mrs. Hofland, with her youngest born on her lap. She has a book in her hand, and is reading aloud to her two oldest children, who have drawn their little chairs close to hers, and gaze earnestly into her face, listening.

Time has dealt gently with Mrs. Hofland. Her clear dark eyes are as bright as when you first looked into them, — nay, brighter, and with a depth of feeling and consciousness not seen before. The fine oval of her face has not changed its curve; the play of feeling is as rapid and rippling; her voice tenderer, deeper, and more musical. You do not think, as you stand looking upon her countenance, over which thought is playing, like sunbeams and shadows that succeed each other rapidly on the bosom of a meadow, that sorrow has been more than once her tearful guest.

She stops reading, and listens. All the children, even baby in the lap, prick up their ears, and look expectancy.

"It's father!" Little Lena is on her feet in an instant, and moving toward the door, with her brother Frank close by her side. Baby Annie's tiny hands are fluttering. In the hall, Lena and Frank spring upon their father, shouting, and hinder the removal of his great coat. But, it is soon laid aside amid these loving hindrances, and the doctor advances to the sitting-room, with an arm around his boy and girl, whose kisses are yet warm upon his lips. Baby's and mother's lips are laid on his as one, making love's circle complete, and sending full currents of joy to his heart.

"A gentleman is waiting for you in the office," said Mrs. Hofland, after the sweet confusion attendant on his return home had subsided.

"Who?"

"He did not give his name. Henry said that he was here before to-day, and asked when you would be home."

Doctor Hofland, expecting to see a patient, or receive a professional call, went down to his office, which was in the basement.

"Doctor Hofland, I believe," said the man, rising.

There was something in his appearance, and the tone in which he spoke, that sent a signal of alarm to the doctor's heart. A shadow, as of approaching evil, fell suddenly around him.

"My name, sir." He hardly knew his own voice.

The man's eyes dropped to the floor, and he bent his head a little forward, as he thrust his hand into his pocket, and drew forth a small bundle of papers. From this he selected a folded document, some nine inches long, by three wide, and said, coldly, as he opened it:

"I have a warrant for you, sir."

"A warrant!" The blood flowed back upon the heart of Doctor Hofland.

"Yes, sir."

"On what account?"

"It is issued at the demand of Warfield."

"Of Henry Warfield!"

"Yes, sir."

"Oh, I'll see him, and make it all right! It's a shame for him to take a step like this. He knows I'll pay him."

"You must go with me to the magistrate's, and give bail for the debt," said the officer, firmly.

The face of Doctor Hofland grew paler. His sensitive pride, as well as his fears, were assailed. He arrested for debt, and required to give bail, or ——! Ah, he knew too well what was beyond the bail-requirement, if not met! The law of imprisonment for debt was on the statute-book of the state, and in active operation, as the full debtor's apartment in the county jail too soberly testified. And creditors, at that time, often made short work with their debtors by forcing them to give security for the payment of their claims, in sight, so to speak, of the jail door. This was the process now taken by one of the doctor's creditors, who had grown impatient and ill-natured.

"But it's night," answered Doctor Hofland. "How am I to get bail at this late hour? The proceeding is an outrage! Who issued this warrant?"

"Mr. Ashmun."

"He knows me very well. Say to him that I will appear and give security to-morrow morning."

"I'm sorry, doctor, but I can't meet your wishes. My warrant requires that I produce your body to-day. I've been here twice before. But you can get bail easily enough. The debt is only ninety-three dollars. Come, if you please."

There was no escape. The hand of the law was on him, and he must stand up, as other men had to stand up, to its full requirement.

"I am called out imperatively," he said, pushing open a little way the door of the apartment where he had left his children a few minutes before. Don't wait tea for me, as I may be detained for some time."

Then the door shut, and Lena heard her husband's feet go quickly down the stairway that led to his office. The tone of voice left echoing in her memory haunted her in a strange way, and troubled her feelings. It had something in it which she did not understand; something that left the impression of a disturbed mind — disturbed from within, not from without — for itself, and not for the peril or extremity of another.

An hour passed, and the doctor did not come back. From the moment of his departure, his wife had felt the pressure of an unusual concern, which continued to increase until vague fears crowded into her heart. After her children were in bed, her mind fell into such an anxious state, that she was unable to read or sew, and wandered about, from room to room, up stairs and down, like a perturbed spirit. Eight, nine o'clock came, and the doctor had not yet returned. But, now, a note from him, so hastily written that she could with difficulty make out the words, was placed in her hands. It read —

"Dear Lena: I shall, I fear, be detained all night. Don't expect me, if I am not home by ten o'clock. Give the children, for me, their go-to-bed kiss.
Lovingly,
E. H."

"Not home to-night! Strange! What can it mean?"

Mrs. Hofland read the note a second time. It's tenor puzzled her. Why did he not say where he was, or hint at the real cause of his absence? This was not like her husband. There was something wrong! What could it be?

And in doubt, questionings, anxiety, and vague fear, Mrs. Hofland passed an almost sleepless night, the first in which her husband had been absent from her since the day of their marrage.

CHAPTER XI.

THE moody silence that followed the scene of strife about money between Mr. and Mrs. Guy, had been prolonged to nearly an hour, when the street door bell was heard to ring loudly.

"Who is it?" asked Mr. Guy, as a servant entered the room where they were sitting.

"A man wants to see you, sir."

"What's his business?"

"He did not say."

Adam Guy, with no pleasing aspect of countenance, for the interruption came upon a scheme of profit half thought out, went into the hall, where he found an ill-looking stranger standing near the vestibule.

"Well, sir?" Adam Guy had already learned the purse-proud art of being rude to persons whom men of his class consider as of little account in the world, except as hewers of wood and drawers of water, and so spoke roughly to the man. Without answering, the visitor handed him a letter.

"What's this?" Guy broke the seal and read —

"DEAR SIR — I am, unfortunately, in the hands of an officer, arrested for a debt of ninety-three dollars, and will go to prison to-night unless I get bail. Will you come to my relief, and save me from this disgrace, and my family from mortification and distress? The bearer will accompany you to the office where I am held. I am grieved to trouble you, but the extremity admits of no alternative.

"Truly

Edward Hofland."

Adam Guy read the letter hastily, and then folding it in a resolute manner thrust it back upon the man, saying coldly —

"I know nothing about it."

"Then you will not go his bail?"

"No, sir! That's a thing I never do. Good evening." And the merchant turned from the messenger, who went muttering from the house.

"Who was it?" asked Mrs. Guy, as her husband returned; but he made no answer. For nearly ten minutes he sat with his chin on his breast — his usual position during the silent evenings spent at home — then, with a curl of the lip, and a veiled pleasure in his tones, he said —

"The Doctor has reached the end of his rope at last."

"Who? What Doctor?"

"Doctor Hofland."

"What about him, Adam?"

"He's in the hands of a constable, and likely to get some experience in jail life."

"What! Oh, Adam! A painful expression contracted the face of Mrs. Guy.

It's nothing more than I've expected. He and his wife began in extravagance and wasteful self-indulgence, and have kept on the same way steadily. No other result could follow. The Doctor has made his bed, now let him lie in it. It will do him good. Men of his class never grow wise until they get a few hard knocks. A short term on the other side of Jones's Falls will make him a wiser and a better man."

"Oh, Adam! How can you talk so coldly!" said Mrs. Guy, showing still greater distress of mind. "Pray go to his rescue! Don't let an old friend be dealt with so cruelly. What is the debt?"

"I made a vow ten years ago, and by my life I'll keep it!" was the emphatic answer; "a vow never to endorse or be security for any man. If my own brother were in Doctor Hofland's place, I'd say as I do now — 'He's made his bed, let him lie in it!' Men like him waste their substance, and run in debt; and then, debt penalties lash them into something like prudence and honesty. I don't pity him in the least."

"Oh, Adam! Adam! Think of his wife and children!" Mrs. Guy wrung her hands, as she looked at her husband through pleading eyes.

"His wife is as much to blame as himself. Lena was idle and extravagant from the beginning," was the hard reply. "Let her feel some of the consequences of her own folly; it will make her a better woman, I trust. No, no; the causes of this trouble are with themselves, and with themselves must rest the consequences. *I* shall not help them, if the Doctor rots in jail."

A shudder ran through the frame of Mrs. Guy, and she threw up her hands in half instinctive horror, as if a monster were before her.

"You needn't whimper to me, Madam, nor put on any of your distressed looks," said Adam Guy, coldly and cruelly, as his wife essayed once more to reach him. "The Doctor's path and mine diverged years ago, and will never run side by side, nor cross again. I want nothing from him, and he will get nothing from me. If he bids for the jail or the almshouse, in heaven's name let him take his election; I wont put a feather in his way."

Mrs. Guy, seeing that no impression could be made on her husband, and pained beyond endurance by the thought of Lena's distress — old, tender feelings were rushing back upon her heart for Lena, between whom and herself circumstances, not alienations, had interposed barriers difficult to pass — left the room and went to her chamber. All her sympathies were quickened into life — sympathies, which contact with sordidness, narrow self-seeking, and hard fighting with an enemy that always seemed on the eve of victory, had only palsied, not destroyed — and she was moved by an irrepressible desire to go to her friend, and offer comforting words, even if she had no power to give aid in her extremity. Hope in her husband there was none. She knew that what he had said he would not gainsay. In all his denunciations of Doctor Hofland, there lay, only half concealed in his tones, a cruel pleasure.

"Poor Lena! poor Lena!" she sobbed, as her pitying heart ran over through her eyes.

"If I had power to aid you, Heaven knows how speedily help would come."

Then, after weeping for awhile, she said resolutely,

"I must go to Lena in her dreadful extremity. I must know all about this trouble in which the Doctor is involved. If I cannot help them with money, I may help by sympathy or suggestion."

Hastily putting on her bonnet and cloak, Mrs. Guy left the chamber, and was coming lightly down stairs, when she met her husband, through whose mind had passed a suspicion of just this course on the part of his wife.

"Where are you going?" he asked, knitting his brows.

"Out for a little while," Lydia answered.

"Where?"

"No matter. I shall not be gone long."

"Going out, and alone, at this time of night! I think it does matter. Answer plainly, can't you? A husband has some right to question as to where his wife goes at an hour like this."

"I am going to see my friend Lena, if you must know." Mrs. Guy looked unflinchingly at her husband, and spoke like a woman in earnest.

"And I say you are not."

"Adam Guy!"

"Lydia Guy!"

Defiantly they gazed in each other's faces for several moments.

"You must not go, Lydia,"

"Why?"

"Let them alone. They have only themselves to blame. Lena is as criminal as her husband, and as deserving of a just punishment."

"Criminal, Adam?"

"Yes, criminal! Havn't they been living on other people's means, and that without consent? Does a thief do more? The law has laid its hard grip on them, and I say it is well. The law is no respecter of persons; they who violate it must take the penalty. I would not interpose a feather to hinder its free course: no, not a feather, in any case. Not in the case of my best friend, even. Let the trespasser be punished; it is our only social safety. Let Doctor Hofland be punished, I say. If he will wrong other people, let him feel the lash. Go back to your room, and don't play the weak fool; the matter is no concern of yours."

"It does concern me, Adam, that a dear friend is in trouble, and, right or wrong, I must go to Lena," answered Mrs. Guy.

"You shall not go! There! I have said it again, and by all that is sacred I will keep my word!" and striding to the door, Guy locked it, and drew out the key. "Now, Madam!" There was a hard, cruel look in his eyes, as he turned to Lydia.

Poor woman! She was not strong enough for open contention with a nature like this man's. She would have gone against his will, and braved the after consequences, if she could have been free of present obstruction; but, face to face with his iron resolution, she found herself like a reed in the wind.

"Cruel of heart!" Lydia moaned out the words in

a bitter wail, as, covering her face with her hands, she sunk upon the stairs.

Adam walked two or three times the full length of the hall, in unusual disturbance of manner; then stopping before his wife, he said, "Lydia!"

But she gave no response.

"Lydia! Do you hear me?"

She crouched on the stairs, her face hidden in her hands, still and statue-like.

"Lydia, I say!" He stamped his foot in out-leaping passion; but she stirred not, spoke not. A shade of concern swept over his face, as he stood looking at her motionless figure.

"Come, come child! this is weak folly — get up!" He had stepped across the hall, and laid a hand upon her arm. A great change was apparent in his voice; it was soft with persuasion. But there came no response. The arm was nerveless, and offered no resistance.

"Lydia!" Something like alarm was now manifest. He lifted her face; it was white! and the dark fringe of her lashes lay as still as if penciled above her cheeks.

"Good heavens! Lydia! Child! Lydia! What ails you? Are you sick?"

As he tried to raise her up, the nerveless form slid from his arms, and he caught it back with an eager grasp, just preventing its heavy fall upon the passage floor. Lifting the fragile body — how light it was! — he bore it to a chamber above. Cold water dashed in the face; friction of the hands, feet and limbs; with other rapidly succeeding means of restoration, gave motion in time to the impeded life-circle, and the pulses began again their feeble beat.

"Poor Lena!" Her heart was still with her old friend. They were the first words that parted her pale lips, in returning consciousness.

Poor Lydia, rather! If Lena had come into a piteous strait, how much more piteous the strait of Lydia! It was, in the one case, but as the falling of a summer storm, or the closing of a summer day; the storm would pass and leave the sky clearer than before — the night give place to morning. But, for Lydia, the sky was leaden with perpetual clouds and unceasing rain: for Lydia, it was Arctic night and winter! The sun of earthly love went down long ago, never to bless her eyes in reappearing. Her path was in darkness, and she must grope on painfully to the end.

CHAPTER XII.

ON his way, in custody of an officer, to the magistrate's, Doctor Hofland ran over, in his mind, a number of persons to whom, in his trouble, he might venture to apply for bail. At first thought, he felt the assurance of a troop of friends. But, doubts obtruded as to one, and pride shrunk from the humiliation of an application to another; so, that by the time he was at the office, he was in a state of painful confusion of mind.

"Will you give bail?" asked the magistrate, after rendering a judgment; for the account had been sworn to, and the Doctor, besides, acknowledged its correctness.

"Of course I will; but such matters are difficult to arrange at night. In the morning, I will bring my security."

"It must be had to-night, sir." The officer spoke to the magistrate. "My instructions are positive."

"Who gave them?" The Doctor turned sharply upon the officer.

"The plaintiff gave them, and we have no discretion."

"The Doctor is an honorable man, Thomas," said the magistrate, interposing.

"I don't doubt that, sir. But I'm a sworn officer, and have no discretion. I must hold his body until the money or the security comes."

"I'm sorry, Doctor. You will have to produce your bail to-night," said the magistrate.

"But, how am I to do that? You held me in custody. I cannot go for a friend."

"Perhaps I can get you a messenger. Harland," the magistrate spoke to a constable, who sat listening with an air of indifference.

The man got up, and came forward.

"Will you take word for Doctor Hofland?"

"If he pays me, I will," was bluntly replied.

"Of course I'll pay you," said the Doctor, with hardly concealed impatience. "How much do you want?"

"I'll go for a dollar."

Doctor Hofland drew out his purse. "There," and a silver dollar passed to the constable's outstretched hand. Now came pause, debate, and hesitation, on the Doctor's part. To whom should he apply? He had many acquaintances, and many friends. A dozen men, whom he felt sure would spring to his relief, the instant they knew of his condition, were thought of; but, in narrowing down the application for security to one after another of these, certain considerations were presented that made his thoughts turn back, in sickening reluctance upon himself. Oh, the bitter humiliation of that hour! Its painful memories went with him to his grave. At last, a selection was made, and a brief note penned hurriedly. It was addressed to a young man of no means, but kind hearted, and an attached friend. He called on him, because he could rely on his friendship and discretion.

Help as well as secresy were needed. To have the thing bruited over the town would be discreditable, and touch his professional standing.

For over an hour Doctor Hofland waited, in the keenest suspense, the return of his messenger. At the expiration of this time, he came back alone.

"Did you see the Doctor?" was asked in an anxious voice.

"No, sir. He wasn't at home; and they didn't know when he'd be in."

"Did you leave my note?"

The letter was handed to the disappointed prisoner who, crumpling it in his hand, walked the office floor for some time in an agitated manner. Then sitting down, he addressed another friend, in trying to communicate with whom, a second half hour was lost. This application gave no better result; the friend was absent, and not expected to return until a late hour.

It was now past nine o'clock, and the officer who had the doctor in charge began to exhibit impatience, and to mutter half incoherent sentences, enough of which reached the ears of Doctor Hofland, to sting his pride and manliness into an agony of pain. The prospect of having to spend a night in jail looked threatening. The gloomy prison stood a mile away from the office in which he was held, and the constable plainly intimated that he could wait no longer, at so late an hour, on the uncertain issue of bail.

"I cannot go to prison!" the Doctor exclaimed, in excitement. "I have scores of friends, who, if they only knew of this extremity, would hurry to my relief. I am well known to you, sir," addressing the magistrate. "There is no risk, as you can assure the officer, in giving me until to-morrow morning to get security. I pledge him my honor, to have a bondsman or the money for which I have been sued, in the office by nine o'clock. This taking a man at fault, in this way, is not fair and right."

The magistrate turned to the officer and added a word in favor of the Doctor, but that official's countenance was hard as iron, and resolute.

"I have no discretion," was his unyielding answer. "The money or bail must come. My instructions are explicit."

"I will make one more effort," said the Doctor, forced into calmness; and he sat down and wrote to his old friend, Adam Guy. It cost him a hard struggle to do so; but pride, and an almost unconquerable reluctance to expose himself in this direction, had to be overcome. He did not doubt for an instant the result, if the merchant should be found at home, and the probabilities were in favor of that. The risk was small, and Guy could not, in very shame, refuse help in such an extremity. A hurried note to Mrs. Hofland was penned at the same time, that she might be forewarned, in case the dreaded imprisonment should result.

Suspense had now become almost unendurable. The Doctor walked the office floor, with the restless, short, quick turns of a caged animal, unceasingly, until his messenger came back.

"Did you see him?" The officer had come in alone. Doctor Hofland's face was working all over.

"Yes."

"What did he say?"

"*No*, point blank!"

"Adam Guy said no!" Surprise and incredulity were in his voice.

"He did."

"You gave him my letter?"

"Yes, sir, and his answer was, 'I don't know anything about it.' Then I said, 'will you not go his bail?' and I thought he'd have taken my head off."

"What were his precise words?" asked the Doctor, now speaking calmly.

"His precise words were, 'That is what I never do. Good evening!' And then he turned from me as if I were a dog."

"Did you deliver the letter I gave you for Mrs. Hofland?" asked the Doctor, his voice faltering a little.

"I did."

"Nothing more can be done to-night. I am ready to go with you." The Doctor spoke firmly as he looked towards the officer who had him in charge. "It's a cruel outrage," he added, "and one of which Henry Warfield will repent."

"It would have been better," remarked the magistrate, "if you had sent notes to several of your friends at once. Ere this time, one or more of them would have arrived. Before going with the officer, I would suggest your writing to one or two gentlemen of your acquaintance, in order that you may be relieved in the morning. Harland will see that the notes are delivered to-night."

"You left the second letter?" Doctor Hofland turned to the constable, named Harland.

"Yes."

"That will do. If the friend to whom it was sent had been at home, I would not be here now. He will make all right at the earliest possible time to-morrow morning."

"He may be here yet," said the magistrate, who was reluctant to see the Doctor so needlessly committed to prison. He drew out his watch, and the officer who made the arrest did the same. The latter shook his head, saying —

"It's a quarter to ten now. I can wait no longer. The jail doesn't stand next door. Come, sir."

Dropping his head upon his bosom, the unhappy prisoner moved toward the door, and passed out silently.

CHAPTER XIII.

Y day dawn Mrs. Hofland is up and waiting expectantly. As footfalls begin to sound along the pavement, she listens for the well known tread of her husband's feet, but listens in vain. One after another, the passers come and go; the number steadily increasing as the day opens brighter and broader. Breakfast has been ready for half an hour, but the Doctor is away still. What can it mean? Lena's overstrained feelings are getting the mastery. The wearying doubts that have perplexed her through the night, have changed to cloudy fears. Some evil must have befallen her husband!

It is nine o'clock, and still no word, no appearance. Office patients have arrived and departed; some still linger, on the assurance that the Doctor is expected to come in every moment. Half past nine. Poor Lena! suspense has become agony. Ten o'clock. The two elder children have gone to school, and she is sitting with the baby on her lap, when the door opens, and the face of her husband looks in. No wonder she starts and cries out in mingled gladness and pain — gladness that her husband has returned; pain in beholding the change wrought on him since his sudden departure last evening. His unshaven face is pale and exhausted; his hair in disorder; his eyes sad and troubled; his garments soiled.

"Oh my husband! Where have you been? What ails you? What has happened?"

These sentences leap from Lena's lips, as she lays her babe down hurriedly, and starts forward to meet her husband. He catches both her hands, grasping them with a close, nervous grip; and, as he holds them, says in a voice that chokes the words, spite of all efforts to speak evenly —

"I've been in jail, Lena!"

"Edward! No — no!"

"Yes, Lena." The voice is steady now — manhood, in a strong, quick struggle, has triumphed.

"In jail!"

"Yes, darling, in jail for debt. It was an outrage."

"For debt! What debt?" Tears are running over her face.

"A debt of some ninety dollars to Henry Warfield. He took a mean and cruel advantage. It was after night when the officer arrested me, and I found it impossible to arrange security at so late an hour."

Mrs. Hofland laid her face upon her husband's breast, and sobbed violently.

"Oh, my husband! My precious husband! That *you* should have been so disgraced! In jail! I cannot bear this!"

The Doctor drew his arm around Lena, and as they passed up to their chamber, he said —

"The lesson may have been needed, dear. I had time to think last night."

"Needed? Oh, Edward!"

"It is wrong to be in debt — wrong for us, I mean," said the Doctor, as he sat down, on passing into the chamber; "we should not have lived beyond our income."

"Deeper shadows fell over the face of Mrs. Hofland. A pang of self-reproach shot through her heart. "It is no fault of yours; I only am to blame," continued the Doctor, who saw into her thoughts. "I have not been a wise and prudent man — have not restricted want to means; and here is the result. How blind — how foolish — how criminal I have been!"

"Don't, don't, Edward! I cannot bear to hear you say this now," said Lena.

"It is wisest to look truth in the face," was answered. "She has been sitting beside me all night, stern of aspect, and I have grown familiar enough with her presence to endure it for awhile longer. She turned the leaves of my book of life backwards, and showed me a record, the reading of which made my cheek red with shame and humiliation. Ah, my wife! there is another law for a man's government in this world, than the law of mere desire. Covetousness is idolatry!"

Mrs. Hofland gazed, in questioning surprise, at her husband. He went on.

"With me, taste and desire have too often ruled instead of prudence; and now, with costly pictures and the like, filling our rooms, I am in debt and at the mercy of eager creditors. This is wrong — all wrong, Lena. Let us begin again — even at the very beginning. This day, I am at least three thousand dollars in debt; and to-night, if a creditor chose, he may send me again to prison."

Mrs. Hofland shuddered, and her pale face grew paler.

"Oh, Edward! Don't say that," she sobbed, tears flowing anew.

"It is the simple, hard truth of the case, dear; and there is no use in disguise," said the Doctor. "The more steadily we look it in the face, the better shall we be able to comprehend our exact position, and the more certainly devise our way of escape."

"Do you see a way of escape?" asked Mrs. Hofland.

"Yes."

"In what direction?"

"The way will be rough, dear."

"No matter. If your feet are strong enough, mine shall not falter. Point out the way, dear husband! Or, turn into it, and you shall find me a brave and cheerful walker by your side."

"I said, we must begin again — even at the very beginning, Lena."

"We cannot do that. The past is past. But, we may change our course."

"We may begin a new order of things."

"Yes."

"And that is what I mean. But, before this is fairly possible, some steps must be retraced. As I sat waiting on the slow moving hours, last night, and watching for the day-dawn, I went over all our affairs, and got at the exact result. It stands thus. The cost of pictures, statuettes, bronze figures, rare and elegant books, coins, medals, minerals, and other things not absolutely required for household comfort, has reached the sum of twenty-five hundred dollars. I propose to sell these by auction. If they bring fifteen hundred dollars only, that will lift half the burden of our debt at once. I feel assured, if the thing is rightly managed, of realizing nearly their cost. I shall arrange in this way;— Have them removed to a room, engaged for the purpose, and minutely catalogued and described. Auctioneers understand the management of such matters. Through advertisements and the distribution of catalogues among the right persons, a company may be assembled that will bid up most of the articles to their cost value. In that case we would be almost freed from debt in an hour. But, this is anticipating too much."

"To sell at auction will certainly involve a heavy sacrifice," said Lena, her countenance not fully responding to the hopeful light which had begun to glow in that of her husband.

"We must expect such a result, and so prepare for disappointment," replied the Doctor.

"A disappointment that will still leave on us the burden of debt."

"But a lighter burden."

"The smallest burden will be as a mountain hereafter," said Lena, despondingly.

"My thought went further," remarked the Doctor, looking steadily at his wife.

"How much further? Did it reach to the entire extinguishment of this debt?" She bent eagerly towards him. "Nothing less than that, Edward."

"It did, Lena."

"Then say on."

"The rent of this house is four hundred and fifty dollars a year."

"Yes."

"Too much for us to pay, under present circumstances."

"I have always thought the rent too high," said Lena.

"We have been no happier here than we were in that cosy nest at first called home."

"Not so happy, I have sometimes thought," replied Lena. "There has been more care for appearances, here; more looking out upon the world; more consciousness of being under the eye of society — and these things take away the mind's tranquillity."

"That dear little house is for rent again. I saw the bill up yesterday."

"Then we will go back to it again. Two hundred and fifty dollars saved in our expenses will, of itself, extinguish a thousand dollars of debt in four years, if no quicker means can be found. But, the change to that house will help more than the saving in rent. Two servants are absolutely necessary in this one; in that I can do with a single servant. This will make a difference of at least a hundred and fifty dollars more in our expenses."

"But, the house will afford no office," said the Doctor. "I've thought it over, but can't settle this point."

"The parlor must serve for an office," was answered.

"Then we shall have no parlor — no room in which to receive our friends."

Mrs. Hofland was thinking rapidly. Where there is a will there is a way, and she found the adage true.

"We can take the room over the parlor," she replied. "There are two rooms, beside this one, on the second floor, and these will give the chambers we need for ourselves and the children."

"There will be no spare room for a friend," objected the Doctor.

"A sofa bed in the parlor can be used on an emergency. But, at present, Edward, only the question of right and duty is before us, and we must settle that, irrespective of other considerations."

"We have twice the quantity of furniture that will be needed," said the Doctor.

"The rest can be sold," was Lena's prompt answer. "A few hundred dollars more will be gained in this way, and debt still further diminished. Out of debt, out of danger, dear husband! Let us act promptly. I shall never have one hour of undisturbed peace, while a dollar of debt remains."

"Nor I; and as peace of mind is, beyond all mere external things, most to be desired, we will seek it in the directest way. Ah, to think what this burden of debt has cost me! What hours of discouragement — what painful humiliations — what a stinging sense of wrong — what fears and tremors. It has robbed me of freedom and manliness. I have felt myself, all the while, in the power of others. It has been the death's head at my feast, Lena."

"But shall be no longer, Edward! Sell everything. I would rather have uncarpeted floors, and the humblest and homeliest things around me, than to see your honor touched or your peace invaded."

CHAPTER XIV.

IKE a true woman, as she was, Lena did not falter. She was stronger in this thing than her husband. The ardor of his purpose cooled, as the excitement of feeling occasioned by that night's imprisonment died away, and he began to look more soberly at the changes proposed. Professional and social pride arose as hindrances. It is easier to go up than to go down. The more Doctor Hofland dwelt on the issues he had looked so bravely in the face at first, the more did he shrink from encountering them. There was, in this receding from the social position he had assumed, an acknowledgment that he had overstepped his means, and been forced back into obscurity. Then taste and love of art, clung to the beautiful objects with which he had surrounded himself. How could he part from these? His rare books and coins, his cabinet of minerals, his objects in natural history — the accumulation of years; must these go also? He could not look this sacrifice bravely and steadily in the face, and said — "It must not be!"

But, Lena did not turn back. That one night of absence from her husband, and the shock that followed when the truth it involved broke painfully on her excited mind, was a trumpet-tongued argument perpetually sounding in her ears. To have the husband she so honored and loved, suffer this cruel humiliation, had dwarfed to insignificance all things else. She would never rest, until he stood beyond the power of any man to lay so much as a finger upon him. Debt must be extinguished, at any and every sacrifice, even to the last farthing — and that in the shortest possible time.

On the evening that closed this day, the Doctor and his wife sat alone in consultation.

"I went past the old house this afternoon," said the Doctor.

"Is the bill up?" There was anxiety in the voice of Lena. A fear lest the house had been taken, crossed her mind.

"Yes."

"Did you see the landlord?"

"No."

"Why not? Some one else may secure it."

"The house looked very small, Lena." And the Doctor sighed faintly, as he let his eyes wander around the room in which they were sitting, and from thence into the one adjoining.

"It was large enough to hold us once, Edward, and is large enough to do it again;" said Lena, firmly.

"We have more in the family, now," rejoined the doctor.

"We are in debt," said Lena, with an emphasis that put nerve into the Doctor's failing heart. "That argument overrides all others. This morning, we decided our course of action; and now, let there be no faltering. You said that we must enter a new way; and I answered yea, and amen! My steps shall not linger, Edward. You pictured it as a hard and difficult way. I see it as plain and easy. That in which we now tread, is hard and difficult. Every step is among hindrances and entanglements. Already there has been stumbling and falling — wounds and bruises — pains and humiliations. We must return, and get upon a smoother, an easier, and a safer road."

"You are a brave, true woman, Lena," said her husband, as warmth came back into his face. "But this going down so far must be prevented if possible. I have been looking over my bills, and find nearly two thousand dollars uncollected, on my books. One thousand of this ought to be realized within three months. I will see my collector, and confer with him in regard to an earnest pressure for settlements. A number of accounts against persons really able, but indifferent as to payment, could be sued out. A thousand dollars within three months would more than satisfy all demands against me likely to become troublesome. My practice is steadily increasing, as you know, and may yield enough beyond our expenses to liquidate everything in a year or two."

"A year or two! Oh Edward! A year or two of debt and danger! No — no; not a month, say I! We thought it out all rightly this morning. Let us be just to others, and just to ourselves. Make out an inventory at once, including every article not absolutely needed, and sell to the best advantage. If you can collect a thousand dollars in three months, so much the better; but don't sue anybody — people may appear less able than they are — forgive, but don't sue. 'I wouldn't have you gain a dollar through constraint of any one. Sickness often impoverishes the means, while it adds to the expenses. Let us command our own resources, and limit our own wants. This is the right way, husband; and the right way is always the safest and the best way."

And so Lena brought him back to his first best resolution. On the next day, the little house, with its sweet garden in front, was secured. Rose-bushes, which they had planted, climbed now to the upper windows, hanging green wreaths, flower-starred, above and around them. A pair of dwarf evergreens, also planted by them, nearly ten years before, stood just within, and on each side of the gate, their graceful top branches bowing in the summer airs a seeming conscious welcome. Running back from these, and grown wonderfully, like children seen after the absence of a few years, stood on each side of the walk leading to the house, a bank of dark green box. Standard roses, and fragrant honeysuckle, towered above less ambitious plants and flowers, filling the air with sweetness.

"How familiar and home-like," said Lena, as she passed through the white gate with her husband, to look at the house, preparatory to deciding on a removal. "Here is the very Champney I planted with my own hands! See how it has grown. And there is the white jessamine I left with its slender arms not six feet high, hanging now its graceful drapery around the upper windows. What a dear, sweet little spot it is, Edward. We shall be happy here again."

The Doctor unlocked the door, and they went in. "The rooms are very small," he said, with a sigh, after pushing open the window shutters.

"They are large enough for love and duty," answered Lena, with a smile. "Better be straitened as to the body, than the mind. These smaller rooms shall be as wide palaces for our freer souls. Ah, husband, dear! — it is from within that our truest pleasures come. If right with ourselves and the world, we may be happy in the humblest cottage. But if not, no princely mansion can give peace to our souls."

They went from room to room, with old emotions coming back into their hearts, and filling them with tenderness. In this chamber, a babe was born; in that one, a babe took its departure for Heaven. Ah, the blessedness and the pain! The joy of birth, and the pangs of bereavement. Softened and sanctified by time and discipline, the joy and the sorrow were felt again.

It went hard with the Doctor to consent at the final moment of decision. Pride and principle came into strong conflict, and but for the unwavering spirit of his wife, he would have receded. But, once fully comprehending their situation — the Doctor had until now concealed from her the extent of his embarrassments — all hesitation, and looking to the right or the left, were gone. Back, back, quickly, from a position of danger! So her heart and her thought said; and as she felt and thought, so was she prepared to act.

Notwithstanding many suggested changes in the programme at first presented by Dr. Hofland, and then approved by himself and his wife, that programme, thanks to Lena's firmness, was strictly carried out; and when they were fairly domiciled in their humbler home, it would have been a hard and selfish creditor indeed, who complained of superfluity. Not a picture nor a book, outside of the Doctor's medical library, that was of value, nor anything merely ornamental, or that could be spared in housekeeping, was removed to the new abode. All were sold: and with what result? Let us see.

CHAPTER XV.

IT was just one week after Doctor Hofland and his wife had taken their step downward, as to external things, but upward, as to the internal. They were alone, sitting in the plain little room on the second floor, which they now called their parlor. The mental discipline, humiliations and anxieties through which they had passed, left on each the sober hues of thought. But, there was nothing of unhappiness — nothing of complaint visible on their countenances.

"I received an account of sales, to-day," said the Doctor, as he laid a folded paper on the table.

"Did you?" Expectation lit up the countenance of Mrs. Hofland — expectation, in which suspense, and a shade of anxiety, were visible.

"Yes, and the result is better than I had any good reason to anticipate."

"Oh Edward! What a relief!" Tears glistened in Lena's eyes.

The Doctor opened the paper, and running down his eyes to the last footing of a series of long columns of figures, said —

"The sum realized is twenty-seven hundred and eighty-one dollars; within two hundred dollars of all I owe."

"My dear husband! I am happier this hour than I have been for years!" Drops of gladness fell over Lena's cheeks. "Thank God for showing us the right path, and for giving us courage to walk in it!"

"Thank God, I say, for so brave, so true, so self-denying a wife!" responded the Doctor, as he caught Lena's hand and pressed it against his heart, where her head was lying a moment afterwards. "I was not strong enough, standing alone, for this," he added. "If you had faltered, our feet would still be in difficult ways — our sky clouded — our hearts in trouble. But now, there is no longer any fear. The way is plain before us. The sky is sunny. I can lift a brave head — I can look every man I meet steadily in the face. Oh, freedom! freedom! It is worth any struggle — any sacrifice. What joy is there in a large house; in pictures; in costly furniture; in the possession of rare books, the leaves of which are not turned once in a year; in gloss and ornament, if a nightmare of debt lies ever on the constricted bosom? How blind, how weak, how irrational I have been? I wonder and am ashamed of myself."

"The lesson is for all time," said Mrs. Hofland, smiling through tears of gladness, that still trembled in her eyes. "We shall not make this error again."

"Never again, Lena!" answered her husband. "What with one hand we take from the world, shall be paid for by the other. If our means are small, we will restrict our wants. Debt shall be an unknown element in our home economy. As for things of taste and ornament — now departed — they will be restored in time, and speak to our souls a higher and truer language than before. This discipline and self-denial, if rightly borne, will open our minds more interiorly, and give them a truer knowledge of the use that lies in the beautiful. Hitherto, a covetous desire to possess has depraved, with me, all love of art; and so robbed me of the higher delights it might have given. I see this clearly, and must strive against and overcome that evil in the mind which has been pronounced idolatry."

"And so," said Lena, "we are not really going down, but ascending in life. This change of position, is not a fall, but a rise. If we see in a clearer atmosphere, and have a more extended vision, we must be at a higher elevation than before."

"We are, Lena. Our embarrassing relations with the world were as clogs, holding us down. The soul sat, groveling, among the meaner things of life; its vision clouded, its strength impaired. Thought dwelt more in the outward than the inward — in customs, usages, appearances, opinions and the like. But, in acting as we have done, from a principle of right and justice, we have emancipated ourselves. The thought of how this and that will appear, is removed, and questions of right or wrong must now determinate our actions. This is freedom; this is growth; this is the soul's true order of existence."

So they talked concerning their newly assumed relation to the world; and while they thus talked, this new relation formed the theme of remark in another household. Let us pass to that of Adam Guy, the merchant.

"Our fast friends have gone over the precipice, as I predicted long ago." There was a gleam of satisfaction in Adam's cold eyes as he thus spoke to his wife.

"To whom do you refer?" asked Mrs. Guy, rousing herself from a state of moody discontent in which she had been sitting for some time.

"Doctor Hofland and his wife." There was as much pleasure in his voice as in his eyes.

"What of them?" Mrs. Guy was all interest now.

"The Doctor had a night's experience in jail a week or two ago."

"More the shame for you!" was answered caustically. "I never could have believed that of Adam Guy."

"Believed what?"

"That you would have abandoned an old friend in such an extremity. Ninety dollars! It will be remembered against you!"

"Indeed!" Spoken contemptuously. "And by whom?"

"People lay up these things."

"People! Pah! What do I care for people, one half of whom I can buy and sell?" And Guy snapped his fingers scornfully.

"What were you going to say about the Hoflands?" asked Lydia, a feeling of disgust hindering any further remark in the direction her husband's thoughts were moving.

"I said, they had gone over the precipice at last;

and no one cares, I reckon. People of their style don't make many substantial friends."

"Why don't they?"

"Fast living and fast friendship are imcompatible things. Your eternal borrower wears out his welcome. You sit uneasily beside a friend whose thought is on your purse, rather than on the theme in which he affects an interest. I know. But the Doctor has found his level at last, and I'm glad of it."

"What has happened to him?"

"You remember that little bird box in which they first lived?"

"Yes."

"His sign is on the door again."

"Doctor Hofland's?"

"Doctor Hofland's. I passed there to-day, and read it with my own eyes. People who stand too high, are apt to fall. I saw, long ago, what the end would be. That night in jail did the work for him, I've no doubt. Creditors are a scary kind of people; when one of their number pounces down on a poor unfortunate, they are apt to follow on swift wings, so as to be in at the death. They've made short work with the Doctor; that's plain. Ha! ha! How it must have surprised him! Well. Let every tub stand on its own bottom, I say. Doctor Hofland has no more right to live off of other people, than your common pickpocket."

"Don't, don't, Adam! I can't bear to hear you talk so about the Doctor. He may have been imprudent; but to compare him with a common pickpocket, is an outrage."

"There's no difference." Guy spoke in a kind of savage ill-nature. "The Doctor's better education increases his responsibility. Men of his class are the respectable pickpockets of society; and what is more in regard to them, their victims are often so tied hand and foot by friendship, consanguinities, social relations, or sympathies, that resistance is impossible. Your footpad or burglar may be shot down; but these decent-faced robbers hold you gently by the hand, and pour honeyed words into your ears, while they rifle your purse. You understand it all, but can make no resistance. I'm always pleased when society spots them, writing rogue on their backs. It has done so in Hofland's case, and I am glad of it."

Mrs. Guy did not answer, but turned herself partly away from her husband, bending close down over some needlework on which she was employed.

"I don't want you to go there," said Guy, who, after finishing his conclusive declaration against his old friend, waited to hear what answer his wife would make. He knew that she had still a warm side towards Lena and her husband — though, through his management, social intercourse had long ago ceased — and uttered his sweeping condemnations more for the sake of annoying her than anything else. He saw from her manner that he had made no impression whatever against her friends, and that grief at their misfortune was the only sentiment stirring in her heart. Remembering how, on learning the danger which threatened the Doctor a week or two before, she had yielded to the impulse, that, but for his interference, would have borne her with swift feet as a comforter to Lena, he had uttered the brief interdiction at the commencement of this paragraph.

"Go where?" asked Mrs. Guy. Her thin, pale lips, closed tightly as the words left them. Her eyes were steady — her brows knit.

"To Dr. Hofland's." The answer was emphatic. Adam saw down into his wife's thoughts. He was quick-sighted in all that came in opposition to his will or wishes.

"If you choose to desert a friend in misfortune, I shall not." Mrs. Guy's utterance was slow, and her tones resolute. "I am going to call on Lena."

"Indeed you are not." There was a quick, short rattle in the voice of Adam Guy.

"We will bandy no words, Adam. You heard what I said." Mrs. Guy's tone was unfaltering.

"I command you not to go!" Passion swept him away into a brutal violence of manner.

"And I shall disobey your command, because you have no right to lay it on me." Mrs. Guy's color mounted, and her eyes flashed. He had struck the smarting spur too deeply.

"You are my wife, madam!"

"Not your slave, sir!"

They glared at each other for a few moments, in angry defiance.

"Go at your peril," said Guy, in a husky, threatening voice.

"At a thousand perils, I will go!" The poor, weak frame of Mrs. Guy was beginning to tremble under the pressure of excitement; but her spirit was strong. Contempt of her husband's mean, cruel, selfish spirit, more of which was apparent to her in his sentences than any reader can perceive, made her spurn his unwarrantable interdiction, as though it were a child's command. "Content yourself with deserting a friend in trouble; but don't ask me to do the same."

"Silence! I wont have such language." The foot of Adam Guy struck the floor with a quick jar.

"As you please," was answered, and Lydia, who had turned towards her husband, turned herself away again, and bent down once more over her needlework; but her hands trembled so that she could not make the stitches, and so she let them fall idly in her lap.

Money is a great power. Out in the world, and among men, its selfish possessor feels himself to be a little emperor in his sphere. He says to this man, "Go," and he goeth; and to that man, "Come," and he cometh; and few there be who set themselves in opposition to his will. He feels that money has invested him with personal consequence, and that from this comes obedience and complaisance; while the truth is, men flow in with his conceits, his plans, his arbitrary will, even, in the hope of advantage. The man himself is nothing. Abstract the money, and he will be of little more account than a sucked orange. It is at home that these mere money-men find the current of their lives obstructed — here, that baffling winds flutter among the sails of their goodly ships, and bear them back from promised havens. Women and children are not so easily managed; particularly when the rich father and husband, not only withholds too much, but exacts too much. He is dealing outside of his dwelling, with material interests; inside, with human souls. Love of gain, of power, of place — all these are potent ministers on the outside; but, on the inside, "I wont," and "I will," clamor against him with an undying persistence. He is not wise enough to govern these home elements, and so sets them at defiance. Unceasing war is the consequence — war kept up to the very last. The children gird on their armor, and learn to handle sword and spear even from the beginning. As they grow older, they gain skill and strength, and the time comes, always, sure as fate, when the battle turns in their favor. But alas! what wreck, what ruin, what desolation, mark the way, and the final victory is but a final disaster to all!

Great as Mr. Guy found the power of money on the outside, inside of his home, the daily conviction grew upon him that he was losing power. His will, yielded to in the beginning, was often now disputed, the ground being maintained on the part of his wife, with a persistence and success that made him feel bitter against her. In the present contest, he was in opposition to the stronger elements. The misfortunes had come upon Lena's old friend, and this so quickened the sentiment of love, that her husband's opposition only fanned it into a blaze. She must see Lena, and the hand of Adam Guy was not strong enough to hold her back. If she sat with fingers too weak to carry the needle — silent, shrinking, and trembling in nervous exhaustion — her will did not give way for an instant. Her heart was drawing her towards Lena with the old strong impulses, and she meant to go as she had said. Comprehending the height and depth, the length and breadth of consequences, Adam Guy had power to visit on her head, she was ready in this cause to brave them. Many feelings that once writhed in anguish when his foot trampled on them ruthlessly, now gave no response. They were dead — *to him*. The bond which united them was external only. Internally, there was repulsion instead of attraction, and aversion instead of love.

No further word passed between Lydia and her husband during the evening. Guy sat for most of the time with brows drawn down, and mouth shut tightly, musing, scheming, pondering, and miserable as he almost always felt when at home for only at home did he find his will thwarted, and his commands set at naught. Lydia passed the hours as she usually passed them, with busy hands, and oppressed feelings. All the outreaching impulses and wants of her woman's nature, had been crushed back, and lay bruised, broken, and helpless, against her heart, that ached, and ached, with a dull, deep, unmitigated pain. Poor wife! The pleasant children which had been born to hope, in the far away years when life and love threw hues of rosy promise on the future, had long ago passed through fire to the golden moloch set up by her husband, and were dead! Mourning them, her spirit refused to be comforted, but sat, tear-eyed and white-faced, in Rachel-like sorrow. Alas, poor wife! Time can never restore these lost ones. They have faded from the earth, and will return no more.

CHAPTER XVI.

O N the day afterwards, Mrs. Guy called, as she had purposed, to see her old friend. It was a long time since they had met face to face; and over two years since their last exchange of formal visits. Her heart was now full of sympathy, pity, and tender interest. The misfortune of Lena had awakened old feelings that came back upon her like a flood. When she reached the pleasant little house, standing modestly back from the street, in which, years gone by, she had passed many sweet hours with this dear friend, it looked so poor and small in contrast with her own spacious and elegant home, that she could not repress a sigh for Lena, as she entered through the gate and moved down the box-bordered walk leading to the door. Her hand trembled as she raised it to the bell and gave a timid ring.

"Is Mrs. Hofland at home?"

"Yes, ma'am," answered the tidily dressed servant, who admitted her to the Doctor's office.

"Walk up stairs."

"Mrs. Guy hesitated."

"Walk up to the parlor, if you please, ma'am." And

the girl conducted Mrs. Guy along the narrow passage and stairway to the front room in the second story.

"What name shall I say, ma'am?" The servant's manner was cheerful and intelligent. Mrs. Guy handed her a card, and she retired. Nearly five minutes passed before Lena made her appearance, and in that time, Mrs. Guy had opportunity to note each article in the room. How mean and meagre every thing looked. The carpet was faded and threadbare, and the scant furniture plain and out of fashion. Only two small pictures hung on the walls, and they were portraits. A pair of china match boxes, and a small gilt candelabra, composed the mantel ornaments. A pair of painted shades, considerably worn, tempered the light at the windows. How painfully all this contrasted itself in the mind of Mrs. Guy, with the attractive surroundings which, on her last visit, made so pleasant the home of Lena. She remembered the choice books and pictures; the statuettes and objects of taste, innumerable, with which her husband had made beautiful their dwelling. Ah, how sad a fall had come!

In the midst of her reverie, Mrs. Guy heard the footsteps of her friend, and rose to meet her. In the moments of intervening suspense, her heart almost stood still. She had pictured a pale, sad, wasted, and despondent countenance; an almost hopeless being with whom she could weep, but offer few words of comfort.

The door opened. Was that bright face, over which smiles were sporting with each other; those eyes, brimming with a loving welcome; the face and eyes of Lena Hofland? Yes, even so.

"Why, Lydia! This is indeed a pleasure!" and she came forward quickly, grasping the hand of her old friend, and kissing her with a heart-warmth that made the sluggish blood leap in new impulses along her veins.

"Dear Lena!" said Mrs. Guy, as they sat down, side by side, holding tightly each other's hands, "I cannot tell you how deeply this misfortune has touched me. I only heard of it last night, and it put sleep far from me."

"What misfortune, dear?" The sober hue that fell over the countenance of Mrs. Hofland, did not by any means extinguish the sunbeams.

Mrs. Guy glanced, meaningly, about the poorly furnished room.

"Oh, yes. I understand you. But, there has been no misfortune, Lydia. This change is wholly voluntary, and marks an ascent, not descent in our fortunes."

Mrs. Guy looked wonderingly into Lena's face. She did not understand her.

"Voluntary, Lena?" she questioned.

"Yes, dear; entirely so."

The eyes of Mrs. Guy went wandering around the room again, and came back to the face of her friend.

"I do not understand it," she said, shaking her head in a grave, doubting way.

"Oh, I can make it all clear. But first put off your bonnet, and lay aside your shawl. You must make me a good visit. It is so long since you were here."

"My heart has been with you, Lena. An old friend is worth a dozen new ones," returned Mrs. Guy, as she drew off her bonnet.

Then they sat down again, side by side, and hand in hand.

"Tell me about this change, Lena. It troubles me," said Mrs. Guy.

And now, the face of Mrs. Hofland grew sober, as thought went back to the painful trials out of which she had just come.

"We were in debt, Lydia," she answered. "Neither the Doctor nor I have looked as closely to the relation between income and outgo, as prudence requires. Our tastes led our thoughts too much away from the homely economies of life, and the result was, embarrassment. Some rough experiences opened our eyes to the wrong and folly of all this, and we made up our minds to go back a little, and make a new start in the world. So, we gave up our house in Charles street, sold off every article that we could do without, paid our debts, and snugged ourselves away in this cosey little place. It was large enough for happiness once, and we still find it so again. The burden of debt being removed, our hearts beat to a lighter measure. No, dear, it was not misfortune that brought us here, but honest independence. If the change works any social alienations, they will not hurt us; for we dwell too much in the real things of life to be affected by any new adjustment of its unreal things. We look more to hearts than faces. To-day has brought me sweet compensation."

Lena paused, looking tenderly into her friend's face —

"What, Lena?"

"Your return, darling." Tears sprang into her eyes.

"My heart has always held you as a precious thing,

Lydia. The old love has never grown dim — cannot grow dim — cannot die. If we have seemed to stand coldly apart, there has been no coldness with me. Circumstance, not interior change, has come between us. I always felt that this was so; and now I know it. To get back an old friend, Lydia, is to gain more than I have lost."

Touched deeply by this, the heart of Lydia gushed in tears from her eyes. She had come, trying in her weakness, to gather up strength to support Lena in the hour of darkness and trial; but Lena was strong, and brave, and cheerful. The storm, which, in her fear, had brought desolation to the heart of an old friend, had swept by without harm. The garden of her mind had not lost a green leaf, nor a fragrant blossom. Before this calm strength, her own spirit bowed in tearful weakness. Strong to comfort, a little while before, she was nerveless now.

"And how is it with you, Lydia?" asked Mrs. Hofland, as she looked more closely at her friend, whose pale, thin face, suggested bad health and a mind ill at ease.

Tears filled the eyes of Lydia again: her lips quivered as she tried to answer. Then she hid her face against Lena, and struggled with the rising tide. A few strong sobs shook her wasted frame.

"Dear friend!" murmured Lena, kissing her forehead, "God comforts; God strengthens."

But, there was no reply.

"It was not good for us to have held apart from each other so long," murmured Lena.

"Oh, no, no, it was not good. But it was my fault, not yours," answered Lydia, "and mine has been the loss. While you have grown strong in the life-battle, I have grown weaker — weaker — weaker. I thought you had suffered misfortune, and came to offer the love and sympathy that was in my heart; but, I find you brave and cheerful. Earthly storms cannot shatter the fair temple your soul has builded — earthly clouds cannot darken its windows, Lena! With you is the beauty of life, with me its desolation!"

"No, no, my friend; do not say that," replied Lena. "There is beauty for all — peace for all."

"Not for me," was sadly responded — "not for me. I have lost my way in the world, and something tells me that I shall never find it again — never."

"Dear Lydia! How strangely you talk. Do not let such thoughts haunt your soul. Tormenting spirits have gained access to your mind and afflict you with their dark suggestions. Look up to God, who is the comforter, the enlightener, the sustainer. He will make a plain way for you: He will strike rifts in the cloud; He will bring you peace."

"Not in this world, Lena." Mrs. Guy raised her head, and turned a pale face, over which a strong calm had fallen, upon her friend. "Not in this world, Lena." She repeated the sentence in a steady voice.

"He will, He will; but you must look up."

"I cannot, Lena."

"Oh, my friend, the promise is to every one. Come unto me all ye that labor and are heavy laden, and I will give you rest. We cannot fall into any state of mind beyond God's reach and sympathy. He came down to man's lowest extremity. We cannot be in any suffering, or darkness, or temptation, through which He did not pass in the Incarnation, and out of which He cannot lift us. He knows our sorrows. He is acquainted with our grief — for into his human consciousness He received all possible human suffering, and by subduing the evil from which it flowed, changed sorrow into joy, and grief into gladness."

"It may be so, Lena; but I have lost my way, and cannot find it again. You have one to lean upon — I stand alone. You have a husband — I am worse than widowed. Dear friend! — bear with me a little, and hear me speak as I never thought to speak in the ear of living mortal. Delicacy, honor, right — all, all, oppose my speech — yet, only in utterance now, can my poor heart be saved from palsy. The sweetness of your life, as I see it now, has made me feel, more painfully, the bitterness of my own. Lena, my soul is imprisoned and starving; and only death can give it release. Adam has shut the door and turned the key."

"Oh, Lydia! Do'nt talk so. I shall think your mind wandering."

A strange gleam shot across Lydia's wan face — a strange light flashed in her eyes. Mrs. Hofland felt a cold shudder run to her heart. The suggestion was unfortunate.

"I should not wonder if it went wholly astray," said Mrs. Guy, mournfully. "Women have lost their reason through lighter suffering than mine."

"This is not well," answered Lena. "Let us be

strong and brave — let us endure and be patient. God's better time will come. Out of much tribulation the saints go upward, at last, white robed and rejoicing."

But Lydia shook her head slowly and sadly, and drawing a little away, said — "If you will not hear me, well. I can keep silent though my heart break."

Instantly Lena threw an arm around her friend. "Dear Lydia! say on. Speak to me as if I were a sister — nay; nearer and dearer than a sister. I hold you in my heart. Your life is precious to me. It is not well with my friend; there is darkness in her soul — her feet are moving along uncertain ways. How is it? Why has the night fallen so soon? Why have her steps wandered?"

"I have no husband, Lena!" The tones struck sharply on the ears of Mrs. Hofland. "There is a man, named Adam Guy who promised to be my husband; a man to whose soul my soul sought to wed itself. But, he has turned from my love and bound himself to another."

"Lydia!" Mrs. Hofland was shocked.

"It is even so, my friend. Human love has died out of him. Gold is his bride."

Mrs. Guy was silent for a time, and then went on. "With Adam, money is the greatest good. Its love has crushed out all other loves. Husband, father, friend, in their true signification — these are no more. Avarice has supplanted them. And I am a woman, Lena; a woman, and bound to this man — hopelessly bound. His wife, in law, and the mother of his children; but of no account in his eyes in comparison with money.

Can a woman bear this? Can a woman's heart beat against a heart of gold, and not be hurt at every pulsation? I tell you no, Lena — no — no — no! There may be those of our sex who, thus conditioned, would compensate or revenge themselves by license, or undying contention; but these are not true women. A true woman must love; rob her of this necessity of her nature, and you darken her whole life, as mine is darkened."

"Dear friend!" said Mrs. Hofland, drawing an arm tightly around Lydia, "you have children. There is mother-love as well as wife-love."

"Children! Yes, I have children!" The tones of Mrs. Guy's voice gave Lena another shock.

"Children!" she continued, bitterly — "Have not the lion's whelps the lion's tooth? — Yes, I have children; or, more truly speaking, a cage of young wild beasts, perpetually struggling against each other, in whom the animal nature grows stronger every day. I grow weaker and weaker, in contention. A little while and they will devour me."

"Lydia, this is dreadful! You are talking wildly. It cannot be so." Mrs. Hofland pushed her friend away, and looked anxiously into her face. She feared the glare of insanity. But, though the eyes of Lydia were tearless and fixed, they gave back intelligent glances.

"I am talking in sober earnest, Lena. It is even as I have said. My children, as they grow older, grow more and more away from my influence. Adam, who is like his father in everything, sets himself against me so resolutely, that I am often powerless in my efforts to move him. If his father is present, an appeal against my authority is generally conclusive. The boy is both avaricious and cruel, and I see these evils gaining strength daily. All that I can do, is like beating the wind. John is forever in contention with Adam, and they are growing to hate each other. Lydia throws herself in mad antagonism against her brothers, and takes more pleasure in strife than anything else. She does not seem to have any moral sense whatever — any conscience — any reverence. And my three younger children are like the elder. I do not wish to live until they become grown up men and women; for they will either tear each other like uncaged beasts, or part in undying hate. Oh, to be the mother of such a brood! Would that I had died a baby in my mother's arms!"

Pent up feelings overflowed their boundaries, and Mrs. Guy fell upon her friend, and wept violently, for a long time.

"Forgive me, Lena," she said, on regaining calmness, "for having intruded things which should have been sacred to myself. I never thought to have spoken thus to any living soul; but, there are times of weakness, when utterance becomes a necessity. Ah, Lena, if I could have talked to you of what was in my heart, years ago, it might have been better. The burden of unexpressed anguish has been too great for me. I am conscious of daily decreasing strength. Mind and body are fast giving way. I feel weak and bewildered nearly all the time. The elements with which I have to contend, are too strong for me."

"God is strong. Lay your burden on him, Lydia."

"I have turned from Him, and He has turned from me," answered Mrs. Guy, in a hopeless kind of utterance.

"Nay, nay, my dear friend! God is an ever present help to all who look to him."

"That may be so, Lena; but we do not look to Him. Ours is a Godless house. No praying; no Bible reading; no church going. We are heathen."

"I do not wonder that you are in darkness and bewilderment, Lydia," said Mrs. Hofland, soberly and impressively, "I do not wonder that your children are growing up in strife. I do not wonder that your eyes look fearfully down the future. If there is no regard for religion in your house; no storing of precious truths from the Bible in the minds of your children; no lifting of hearts upward in prayer to God, the case is bad indeed. You must try to change all this."

But Mrs. Guy shook her head, murmuring, in a weak way — "I cannot."

"Don't say that, Lydia. You can, if you will. If the older children are, as intimated, beyond your influence, begin with the little ones. Save them."

At this moment Mrs. Hofland's two oldest children entered the room, quietly, an arm of each around the other's waist.

"Who are these? Not your Lena and Frank?" said Mrs. Guy, reaching her hands to the children, who came to her side in a respectful way, and looked pleasantly into her face.

"Lena and Frank," replied Mrs. Hofland, as a bright

smile lit up her countenance. "This is Mrs. Guy, don't you remember her?" And she spoke to the children.

Lena said yes, and Frank stood silent, with his looks modestly cast down. Mrs. Guy kissed them, tears filling her eyes as she thought how rudely and boldly her oldest children would have dashed into the room, had she been at home, and Mrs. Hofland the visitor.

Their entrance having interrupted the conversation, when resumed, it kept away from the unhappy subject in which it had dwelt from the beginning, and reached a more cheerful elevation.

"You will come to see me, Lena?" said Mrs. Guy, as she held tightly the hand of Mrs. Hofland, at parting.

"Oh yes."

"Come soon."

"Yes, very soon."

"Remember me to your good husband. I wish he were, as once, Adam's friend."

"He would stand his friend to-day, Lydia, if there were any need of service. If there is a distance between them, it is not, I can assure you, the Doctor's fault."

"I know that, Lena. Adam proved himself unworthy of such a friend. Whatever distance intervenes, he made it. But we will not talk of that. Good by, dear! Come very soon. You don't know how much good it will do me."

There was a prolonged, tightly given pressure of hands, and then the two friends separated. Lydia returned to her large, elegantly furnished house, and to her husband who counted his gold by many thousands; but returned with a heavy heart. It looked, in her thought, more cheerless, more desolate than ever, now that she had felt the love-warmth of Lena's home. She went, in pity and sympathy for an old friend in misfortune, but returned, sadly conscious that with her was the misfortune. and with Lena the sunshine of a true prosperity.

CHAPTER XVII.

THE sad revelation made by Mrs. Guy touching her home-life, wrought a painful impression on the mind of Mrs. Hofland, whose feelings were strongly interested for her old friend, and went out towards her in a yearning desire to give help, comfort, and strength to bear up under the heavy burdens laid upon her weak shoulders. She was in her thought nearly all the while. On the second day after her visit, Lena called on Mrs. Guy. It so happened, that Mr. Guy had returned home for some purpose late in the forenoon, and was leaving the house, as Lena came up the steps. Mrs. Hofland smiled, and said,

"Good morning, Mr. Guy."

The merchant frowned, nodded coldly, and passed her in a rude manner. It was meant for the cut direct. For an instant, Lena hesitated to ring the bell. But a thought of her unhappy friend enabled her to throw the insult behind her as a thing of no account. She found Lydia with eyes wet from recent weeping.

"It's the old story," said Mrs. Guy, answering her questioning looks of Mrs. Hofland, and trying to smile indifferently as she dried her tears, — "The old story of strife about money." And she held up some bank bills that were crumpled in her hand. "I asked Adam, just now, for a hundred dollars; and here are fifty, just half of what I need. It is always so. If I ask for twenty, I get ten, and hard words to make up the balance. I'm the most extravagant woman that ever lived. How did I manage when my whole income came through my needle? — ha! So he talks. Money! Heaven knows, I often wish there was none of it in the world. But, didn't you meet Adam at the door?"

"Yes; but I don't think he recognized me."

"Not recognize you!" Mrs. Guy's countenance changed a little.

"No. He passed me with a distant nod, as if I were a stranger."

The eyes of Lydia fell to the floor, and she sat musing for some time.

"How long is it since you have met him?" she inquired, looking up.

"Nearly three years."

"I don't see that you have changed in anything. But he may have forgotten you. His thought is so fixed on money and business, that it would be no matter of surprise if he forgot the face of one of his own children after an absence of six months."

"How are you?" said Lena, after a pause, seeking to get away from this unpleasant theme.

"About as usual, and that isn't much to boast of. But, I'm really glad to see you, and must ask forgiveness for so cold a welcome. I'm not always able to rally myself in a moment. I wish, sometimes, that I had no more feeling than a stock or a stone; that I didn't care for these things. But, woman's nature is weak. We cannot harden under perpetual blows; but grow more and more sensitive even to the last stroke that extinguishes life. Again, I say, forgive me. The pent up anguish of my spirit found an outlet in the direction of your sympathy, and I cannot close it again. Bear with me, Lena! I know that it pains you to hear me speak as I am speaking, but I cannot, in the fullness of my heart, keep back all utterance."

"Look away from what, in the present, dear Lydia, is irremediable. To bear, is to conquer. What we brood over, gains new vitality. As far as possible, veil even from your own eyes the harder aspect of your way in life, and look forward in hope, to some more pleasant future."

"The future is darker than the present, Lena. But this is all wrong, I know. It isn't kind in me. I shall lose you again, if I worry your mind after this fashion. How weak and unreasonable I have become."

Very much in this strain did Mrs. Guy talk during the visit of Lena; and in parting, she wept bitterly, saying—

"I know you won't come here again. It's so wrong in me; but I've grown weak and childish, and can't help it."

"Come and see me often, Lydia," was the kind answer of Mrs. Hofland, as she kissed her unhappy friend. "I shall hold you always in my heart. Let me be as your sister. Talk to me without reserve, if talking gives any comfort, and what you say shall be sacred between us."

"And you will come to see me, in return."

"Oh, yes, often."

"You are true and good, Lena, and may Heaven bless with richer blessings than even now rest upon your life," said Mrs. Guy, as they parted at the door.

On the return of Adam Guy, at dinner time, his first words on meeting his wife, were —

"What did that fellow's wife want here?"

"I don't understand you," answered Lydia, coldly. "Of whom are you speaking?".

"You know very well of whom I'm speaking."

But Lydia shook her head perversely.

"Wasn't that Dr. Hofland's wife I saw at the door this morning?"

"Lena called to see me; but you didn't mean her when you said that fellow's wife?"

"I meant her, and you know it. What did she want?"

"If you were curious on the subject, you should have inquired yourself," returned Mrs. Guy, with ill-disguised contempt in her tone and manner. "So you knew her?"

"Of course I knew her."

"And passed her without recognition?"

"I did, and mean to always."

"Why?"

"Because I don't like her nor her principles. She's not a true woman, and I warn you to have nothing to do with her."

"Not a true woman! Heaven save the mark! Pray draw a picture of one. I would like to have your ideal above all things. Perhaps I might copy after it."

"Oh, you can sneer! but that amounts to nothing," retorted Guy, rather impotently. His wife's scorn grew sharper every day.

"Look here, Adam," said Lydia, speaking resolutely — "I don't trouble myself in regard to your friendships, and I beg you will not trouble yourself in regard to mine. I have been to see Lena, as I told you, and Lena has returned the visit. It shall be no fault of mine if the restored intercourse is not perpetual."

"Very well, madam. Set yourself in defiance. But don't complain of the consequences. You wanted a hundred dollars this morning. I understand it now."

Lydia, who had been turning away from her husband, wheeled round, under a sudden impulse, and confronting him, with flashing eyes, said —

"What do you mean, sir?"

"I presume you understand me," was replied in a cold, sneering manner. "Where are the fifty dollars I gave you?"

Mrs. Guy thrust her hand into her pocket, and taking therefrom the roll of bank bills received from her husband a few hours before, flung them into his face, saying —

"There they are! Take them again. If your soul is made of money, there are other souls of better material, thank God! Adam Guy! — Doctor Hofland and his wife don't want your money. They are richer

7

than you are, or ever will be, though you live a thousand years, and double your possessions each year."

The money struck the face of Guy, and fell at his feet upon the floor. The act stunned him. There was a look and tone of defiance in his wife that overawed him for a little while. He did not understand the way to deal with this aspect of antagonism.

"Keep your money, if you will, sir!" added the excited and outraged woman. "I hate the name of money. It is an offense to me. From this day, my lips shall not utter the word to you. Dole it out as you may, in miserly pittances, it will be all the same to me. There is not a woman in the city, sir, whose husband's property reaches, at the utmost, half of your possessions, whose wardrobe is not twice the value of mine. I have been ashamed to appear in company; but that feeling is gone. The discredit is yours, not mine."

"Silence, madam! I will not hear this!"

As often before, when he felt himself borne down by his wife's indignant reaction upon outrage, Guy stood upon authority, and commanded silence.

"It wont do, Adam Guy," said Lydia, with a smile curling her pale lip. "You may rob, but you cannot silence me."

"Rob! are you going crazed?"

"Yes, rob; that is the word. He that withholds what is just, is as much a robber as he that plunders by force; and meaner, because more cowardly. Do you understand me?"

"No."

"Turn it over in your thought as often as you turn

a dollar before spending it, and perhaps the meaning will be clear."

"Your precious friend has been giving you some lessons in duty, I see," retorted Guy. "A few more visits, and I'll find the door locked against me. After ruining her own husband, she has become ambitious of more extended operations. I'll send the Doctor a note, requesting him to keep his vicious cattle at home."

"Happily, the Doctor knows your quality, and will take the performance for what it is worth," said Mrs. Guy, nothing daunted by the vulgar threat. "Men who stand at his height, read such as you at a glance. Send the note. It matters nothing to me."

Baffled by the coolness of his wife's scorn, Adam Guy broke out again into passionate command. Lydia fixed her eyes sternly upon him for some moments, holding his gaze long enough to let him understand that she defied him; then, turning from him, she left the room.

At his feet lay the crumpled bank bills, thrown by Lydia in his face a little while before. Most men, after such a scene, would have let them lie on the carpet, if certain of their being swept into the street. But, in his eyes, money was too precious a thing to be left in any jeopardy. So, stooping to the floor, Guy took up the bills, and thrust them into his vest pocket, muttering in an undertone —

"A good illustration of the value *she* sets upon money. A man might as well pour water into a sieve, as place it at the discretion of such a woman."

In spite of the insult he had received from his wife,

Adam Guy felt a secret pleasure growing out of her declaration that she would never again ask him for money. He wished in his heart that she might stand by her threat. There was no way in which she could inflict self-punishment so agreeable to her husband as this. Her demands for money, so incessantly made, and so steadily resisted, he had always regarded as excessive. This had been the bone of contention between them from the beginning. Always doling out reluctantly, and too often, in complaint of extravagance, he had kept Lydia so bare of money, that constant application became a necessity. To-day, it was two or three dollars for a seamstress; to-morrow, a dollar for the washerwoman; the day after, five dollars for market money; and the day after that, a dollar and a half for sawing and putting away a cord of wood, for which the poor wood-sawyer had waited two hours. So the changes rung incessantly. It was literally true, as he often alleged — "Money! money! — nothing but money! The first thing in the morning and the last thing at night. I can't show myself without hearing the word money!"

He would not trust his wife with any large sum for disbursement. We doubt if he ever gave her so much as a hundred dollars at one time in his life. That kind of liberality would, he felt sure, encourage extravagance. He must hold the purse-strings tightly, and know for what use every dollar that left his possession was given. No wonder then, that it was "Money, money — nothing but money." His own act made perpetual demand the sole means of home subsistence.

Was Lydia really in earnest in what she had said? He dwelt on her declaration curiously, even hopefully. No sense of shame touched him. Avarice had long ago smothered shame.

"We shall see!" fell from his lips, as he moved about the room, conscious relief following the words, "We shall see! Home will become a second paradise!"

The dinner bell rang, and Mr. Guy stalked moodily into the dining room. A side-glance at his wife's face, who did not look towards him, revealed an expression of fixed resolve not often seen there. He was a little puzzled. The meal passed in almost dead silence. As for the children, they read in their parents' faces enough of warning to induce orderly conduct. Experience had made them observant; and they knew when trespass would be visited by certain banishment.

As Mr. Guy arose at the conclusion of his hastily eaten meal, he tossed the little roll of bank bills across the table, and without a word, retired.

CHAPTER XVIII.

 A DEAD calm followed this scene of contention between Lydia and her husband. One week, two weeks, glided away, and, sure enough, Adam had not heard the word money issuing from the lips of his wife — nor, in fact, many other words. She moved about, when he was at home, in a silent, gliding, ghost-like way, that struck him as unnatural. When he spoke to her, she usually answered without looking at him. If her eyes rested in his, their expression caused an uneasy feeling to creep through his mind.

"We'll see how long this will last," expressed Adam's thought and purpose. "A thing worth having, is worth asking for." So, money was not offered to Lydia.

One day, early in the third week of this new order of things, as Mr. Guy sat in his counting-room, talking with a merchant on business, a black man came in, and handed him a note.

"Good morning, Abe," said the merchant, recognizing, in a kind way, the black man.

"Good mornin', Massa Williams," returned the negro, respectfully.

"What's this?" asked Mr. Guy, knitting his brows, and speaking sharply. He had opened the note, and read—

"Due Abe for Whitewashing, - - $5.
"Lydia Guy."

"Missus guv it to me, sir. I'se done de whitewashin'."

"Didn't she pay you?" demanded Guy, not clearly understanding what the due-bill meant, and exposing to the merchant-friend more than he found at all pleasant to think about afterwards.

"Oh, no, Massa Guy. She say, take dat to Massa, and he'll pay. The whitewashin's all done fust-rate, Massa Guy!"

"Why didn't you wait until I came home this evening? What did you call here for?" said Mr. Guy, as he drew out his pocket-book. He was excessively annoyed, and had not sufficient control of mind to hide his feelings.

"Missus say, go to de store!" Abe's white teeth glistened, as he stood smiling and apologetic.

The five dollars were paid, and Abe retired; but, scarcely had he passed into the street, when a stout countryman entered, and presented another piece of paper. Mr. Guy caught at it in a nervous way.

"Due John Thomas, $10, for milk and cream.
"Lydia Guy."

"Who told you to bring this here?" asked Guy, roughly.

"Your good lady, sir," replied the man, respectfully.

"Henry, pay this, and take a receipt to date," said Mr. Guy, looking round at the clerk; and he turned from the man with a most ungracious air. But, ere the broken thread of business conversation had been fairly taken up, one of his house-servants entered the counting-room.

"What do you want, Hannah?" said Mr. Guy, knitting his just relaxed brows.

"Mrs. Guy said ye'd give me my money," replied the girl, handing him a folded note. The contents were—

"Due Hannah, one month's wages, - $6

LYDIA GUY."

Couldn't you have waited until I got home?" angrily demanded the merchant!

"No, sir. I'm to send it till Ireland; and it must go the day. I towld her yestherday that I'd want it, and she said, very well. An' to-day she gev me this to bring till yez, sir."

"Outrageous!" muttered Guy to himself. "What does she mean? Then handing the due-bill over his shoulder, he said—

"Henry, pay this, also!" As the girl, after getting her money, was retiring, Guy called out, "Hannah."

"Sir, till yez." The woman's voice was not over respectful.

"Next time you want money, wait until I come home."

"Maybe, if ye didn't keep the mistress so close ——"

"Silence! How dare you!" Guy broke in angrily upon the servant's impudent retort.

"Och! An' yez may scrame silence till thim thot cares; but ye nad'nt thry ut wid me, Musther Guy. The leddy hadn't ony money, and she towld me to come here. No mighty harum done, I reckon."

And with this speech, the free-tongued Irish woman, who had seen enough of Guy in the family to despise him, flung herself out of the counting-room, and made quick exit from the store.

"Well, if that doesn't beat the Old Boy himself!" said Adam Guy, his face flushed with shame and anger. But the play was not over yet. A shabbily dressed boy came shuffling into the counting-room a few minutes afterwards, and standing in front of Mr. Guy, commenced operations on an old pocket in his trousers, whose heterogeneous contents were half removed before the object of his search was found. Guy felt nervous. Was here "another cursed due-bill?" We give the words he uttered in thought. Even so; for scarcely had the question formed itself, when out came a rumpled piece of paper, which the boy held towards him, saying —

"Mother told me to give you this, and you'd pay it!"

"What is it?" Guy caught the slip of paper from the boy's hand, and glanced at the single line written thereon —

"Due Aunty Green, - - - 64 cents.

"LYDIA GUY."

"Here! Take this back to your mother, and don't dare to show your face in my store again." Guy lost his temper completely. This was the last feather.

"Good day," said the merchant with whom he had been in conference. "I'll drop in again, and talk over that matter."

"Good day," was returned, coldly, and the merchant retired. But the boy remained standing, with the due-bill in his hand.

"Didn't I tell you to be off?" And Guy advanced upon the lad with a threatening look. The little fellow, however, stood his ground.

"Go, I say!"

"Mother said you'd pay me sixty-four cents. Mrs. Guy wrote it down on the paper."

"I shall not pay it; so off with you this instant!"

Two angry spots burned on the lad's cheeks, and his eyes flashed like diamonds. Moving back, until he stood in the counting-room door, and in a safe position for retreat, he screamed out—

"Stingy old hunks! Cheat a poor woman out of sixty-four cents!" and then ran off at full speed.

Catching up his hat, Mr. Guy left the store in a hurried manner, and proceeded homeward. Stalking into the room where his wife sat with two or three of the children, he said, in a rough, angry voice—

"What's the meaning of all this?—ha!"

"Meaning of what?" asked Mrs. Guy, without evincing any surprise at her husband's manner.

"You know well enough!" stormed the excited man. "Don't put on that weak pretence!"

Lydia dropped her eyes from his face, and pursued quietly, and with a steady hand, the work on which she was engaged.

"Did you hear me?" The heavy foot of Mr. Guy jarred the floor, as often in times gone by; the effect was the same as if his wife had been a statue. There was no response.

"Lydia!" The voice was pitched to a lower key, and to a different modulation.

"Well." She paused in her work, and looked up.

"Why did you send them people to me for money?"

"It was due them." The dead level of Mrs. Guy's tone and manner baffled her husband.

"Don't do it again! I wont have Tom, Dick and Harry, running to the store after money. I'm surprised at you! And as for Hannah, the insolent huzzy!—can't stay in this house another day."

Mrs. Guy dropped her eyes upon the sewing in her lap, and the needle-hand, which had been suspended in the air, moved on again — stitch, stitch, stitch.

"Why didn't you tell me you were out of money?"

Mrs. Guy gave her husband a look so full of a strange, half-understood significance, that his breath stood still for a moment. Drawing out his purse, and taking therefrom bank bills to the amount of forty dollars, he gave them a twist in his fingers, and then threw them across the room towards his wife. They fell on the floor, several feet from where she was sitting. She did not glance towards them, nor pause in her sewing. Guy, as he tossed her the money, turned away, and left the room.

On the next morning, while Mr. Guy sat with the

same merchant who had witnessed his mortification on the day before, in the midst of a closely driven bargain on both sides, a girl, wearing a sun-bonnet, and having a checked apron over a faded calico dress, came into the counting-room, and said—

"Is Mr. Guy in?"

"That's my name. What do you want?"

The girl opened her hand, in which she held a narrow, folded strip of paper.

"Mrs. Guy told me to give you this, and said you'd pay it."

An angry heart-beat, sent the blood in red stains to the face of Adam Guy. He took the slip of paper, and read—

"Due Mrs. Winter, for butter and eggs, $7,41.

"LYDIA GUY."

"This is beyond endurance! What does the woman mean?" exclaimed Guy, losing command of himself, and betraying, in the sentence, a glimpse of the skeleton that was in his house. Then adding, impatiently, as he looked towards a clerk—

"Pay it, Henry."

"See here, girl!" he said, roughly, as the person who had brought the due-bill was about retiring with the money, "don't bring any more of them things here. My house is the place."

"You needn't be so huffy about it," retorted the girl, whose rough contact with life in the markets had made her quick-tongued and independent. "A body's a right to ask for their own anywhere. Mrs. Guy said come here."

"Off! Off!" And the humiliated merchant waved his hand.

"Highty!" ejaculated the market girl, as she moved back, and glided through the door, "what's to pay now?"

Amused glances passed from clerk to clerk, as they looked after her, retiring, with a jaunty air, through the store. Ten minutes later, and another due bill, for a trifling sum, came in; and before dinner time three more were presented. Guy was boiling over when he reached home at two o'clock, his dining hour.

"What did you do with the money I gave you yesterday?" he demanded, stalking into the presence of his wife, and thus interrogating her before all the children.

"I received none," was the cold, indifferent answer.

"What? I gave you forty dollars yesterday!"

Lydia merely shook her head, and murmured passively,

"You are under a mistake."

"Didn't I throw you some bank bills yesterday, in this very room?"

"Did you?"

"Certainly I did. Where are they?"

"Perhaps you'll find them on the floor, where you threw them; they never came into my possession," was the impassive answer of Lydia.

"What! You don't mean to say that you left forty dollars lying on the floor to be stolen by servants, or swept into the fire?"

"No, I didn't do any thing of the kind. If so foolish an act took place, the folly may lie at your door; it certainly does not at mine."

Circumvented, Adam Guy! This weak woman is proving too strong for you.

"Didn't you see the money when I threw it towards you?"

"Yes."

"Well! Why didn't you take it?'

"I'm neither a dog nor a beggar, Adam Guy! If you wish me to disburse the family expenses, place the means, in a decent way, at my disposal."

"But where are the forty dollars?" Ah! Here was the pinch! And Guy began to look about the floor. "Adam! Did you see anything of the money?" addressing his oldest boy.

"No, sir," was promptly answered; and then, with the eager scent of a hound, this money-loving child began hunting about the room. The sofa was dragged from the wall; edges of the carpet pulled up here and there; tables and chairs moved from their places; and search made even in the ash pan of the grate. But, to no good purpose.

"There's no use in looking," growled the unhappy man. "Of course the money's gone! swept into the fire, or the street. It beats every thing I've yet seen! No more value is placed on money in this house, than if it were so much dirt."

"I've found it!" cried young Adam, who had continued to prowl about, moved by his avaricious instinct, after all the rest had abandoned the search as idle. And he held up the little twisted roll of bills, that, by some strange chance, had lodged in an out-of-the-way corner of the room, behind a piece of furniture.

A stranger would have thought, by the joy which instantly made radiant the face of Guy, that this sum of money was all he possessed in the world. Catching the bills from Adam's hand, he opened and counted them over in an eager, nervous manner.

"Are they all there, father?" asked the boy.

"Yes, my son; fortunately. Such outrageous indifference beats every thing!"

Mrs. Guy had shown no interest in the hurried disorderly search, which had ended in finding the lost bills, and gave no sign of pleasure at their recovery.

"Here!" said her husband, now thrusting the money almost into her face. "Do you see it now?"

But Mrs. Guy did not move a hand.

"Why don't you take it?" was demanded, in a tone of authority.

"I've told you before, that I'm neither a dog nor a beggar, Adam Guy!"

The look that flashed out upon Guy from the suddenly lifted eyes of his wife, caused him to move back a step or two. The voice was cold and steady; but the eyes had a gleam in them that caused a creeping chill to run along his nerves. He stood, holding out the money for a little while, and then, seeing no movement on the part of his wife, gave it a safe lodgment in his pocket-book.

CHAPTER XIX.

AFTER eating his dinner, Mr. Guy arose from the table, and coming round to where his wife sat, laid the money which she had refused to take from his hand beside her plate, saying, in his ungracious way,

"You see that, don't you?"

She did not answer, nor touch the money.

"Lydia!"

"Well, sir?" A cold gleam went up into his face.

"You see that money?"

"I have eyes."

"Oh, well, I'm glad. Then you see the money. Pray, don't let it go into the fire."

"I would suggest the same to you." And Mrs. Guy arose from the table and left the room.

"Did any one ever see the like of that," muttered Guy, in a baffled way, as he caught up the bills.

"She doesn't know the use of money, does she, father?" said young Adam.

"O, dear, no!" responded the father, in a half despairing voice.

"She'd waste and scatter faster than ten men could

earn," added the boy, drawing from his memory a sentence which he had treasured from his father's lips.

"Yes, faster than forty men," was answered, in strange thoughtlessness, or indifference, as to the ears that drank in the words.

Guy went off to his store without seeing his wife again. A little slip of paper, in the hands of a colored man, reading thus —

"Due Jim Lane for oysters, - - $1.40,

"Lydia Guy,"

pricked him sharply during the afternoon, and admonished him to settle this question of money on some basis that would be satisfactory to his wife. The due bill annoyance had come to be a source of amusement with the clerks, who all knew him well enough to dislike and despise him; and more than once he caught their smiling interchange of glances, as the demands came in. The meaning of it all, they were not slow in guessing.

"This has gone far enough, Lydia," he said, when they were alone in the evening.

His wife looked at him without answering — looked at him with a cold indifference of manner.

"I wish you to pay for everything as you get it. No more of those bills and due bills. It must be stopped short off."

No reply

"Do you understand me, Lydia?"

"I'm not certain that I do."

"I said, that you must pay for everything as you get it — no more of these bills and due bills."

"Just as you please. It's a matter of indifference to me." Mrs. Guy's voice was at a dead-level.

Guy gave utterance to a few words of angry impatience, but they provoked no answer from his wife.

"Make me out a statement of expenses, that I may know what sum to supply. I'm sick of this working in the dark — this pouring out of money in an incessant stream, and seeing it disappear like water in the sand. Here's a small blank book. You must keep an account of what you spend. Set down, on this side, all you receive, and on this side, all you pay out. That's the way to do. I've wanted this system from the beginning, and said so a hundred times. Now, I insist upon it."

He reached the book towards Lydia, who took it from his hand, and without apparent feeling, tossed it lightly into the grate where a strong fire was burning. The flames curled eagerly around it, and threw a bright glare over the room. Guy started to his feet, exclaiming in a hot passion,

"Madam! Are you insane!"

Three or four hasty turns were made through the apartment; then the excited merchant stopped before his wife and confronted her. She sat, with her chin drawn down, looking up to him with a cold smile of triumph in her eyes — a smile so singular and unusual, that he shivered under it into calmness.

"What do you mean, Lydia?" The question was in a greatly subdued tone.

"Nothing but self-protection," she answered.

"Self-protection!" Adam Guy's lip curled. "You are playing at a bold game, madam; and will, in all probability, find that you have mistaken your man."

"As you have found, already, that you have mistaken your woman. But, we shall see!"

Her tone was implacable.

Guy endeavored to look his wife out of countenance, but failed. There was a new expression in her eyes that he could not fathom, and a meaning in her air, voice, and conduct, that threw him entirely at fault.

"How much money do you want for expenses? That's the matter in hand, now," he said, recovering himself, and coming back to the theme uppermost in his mind.

"I didn't ask for anything," replied Lydia with irritating indifference.

"Confound it all!" stormed Guy, breaking away from all self-control. "Are you possessed of a devil?"

"Perhaps," his wife answered. And another gleam shot out upon him from her strange eyes.

"Will forty dollars a week supply your wants?" said Guy, taking out his pocket-book. His manner was changed.

"I have no wants," she answered, with provoking indifference.

"Will forty dollars supply the wants of the family, then? You know what I mean."

"Can't say," replied Mrs. Guy.

"Can't you guess?"

She merely shook her head.

"Well, here's fifty. That must serve, surely." And Guy held the money towards his wife. But she did not raise her head.

"Why don't you take it?" he asked.

"I'm neither your slave, nor your dog, nor a beggar, Adam Guy! Can't you understand me?"

Her eyes flashed; her cheeks burned; her pale lips quivered with feeling. Starting to her feet, with the springy bound of an animal, she stood with him face to face, in attitude and expression proudly defiant. He moved back a step or two.

"No, I don't understand you," Guy answered. "All this passes my comprehension."

"I'm sorry for you, then. But you *will* understand me."

"Why don't you take the money?"

"Simply, because it isn't rightly tendered. There's to be no more tossing of your dirty rags in my face, Adam Guy! I'm no beggar to pick up your crumbs; no slave to accept your grudged offerings and be thankful. But your wife and your equal in all things; and as such, I will be treated with respect, if not kindness."

"You will!" Guy was recovering himself. He retorted with a rising sneer.

Lydia raised her hand in a warning way, and sent a glance through and through her husband. He paused and wavered.

"Pray, give a formula, that I may know how to conduct myself." His tone was slightly contemptuous.

"Conduct yourself like a gentleman," was the calm, dignified answer. "That will cover the whole ground. I ask for nothing more, and will accept of nothing less."

A dark scowl settled over the face of Adam Guy. He found it impossible to go any further in the way across which this new obstruction had been thrown, and so stepped back from it; not, however, in weak acceptance of an ultimatum, but to scheme and plot over the means of getting it out of his road. He was too strong-willed — too much in the habit of compassing his ends, to retire from this field. On the next morning, he again tendered money to his wife, saying, now, in a kind, respectful way —

"Here are fifty dollars, Lydia, for expenses."

Mrs. Guy received the money with a quiet air, and placed it in her pocket.

Three days afterwards, a woman who kept a small dry goods store to which Mrs. Guy was in the habit of sending or going for tape, needles, trimmings and the like, called on Mr. Guy at the store, and presented a due bill, signed by his wife, for twenty-seven dollars and a few odd cents. On the same day, the baker dropped in with another due bill, calling for sixteen dollars. Guy paid them both, without a sign of feeling, just as if disbursements in this way were a part of his system. Already there had been sufficient of mortifying exposure in the face of his clerks, and he was not inclined to lift the veil again. But, to have due bills to the amount of over forty dollars presented within three days after giving his wife fifty dollars, struck him as a calamity. This was indeed, he felt, like pouring water on the sand.

"If I were a millionaire, I could not stand this!" he said, in his thought. "The woman is losing her senses."

In the evening, Guy endeavored to approach his wife with remonstrance on the money question, but she pushed him aside with a cold dignity that chafed him into passion.

"Madam!" he exclaimed, "I will not have my goods wasted — my hard accumulations scattered to the wind!"

Lydia made no response; not even so much as lifting her eyes from the book she was reading.

"Where are the fifty dollars I placed in your hands, day before yesterday?"

No answer — no sign.

"Lydia!"

Mrs. Guy looked up.

"Did you hear my question?"

She bowed, indifferently.

"Then why don't you speak?"

"You have got to learn another way with me, Adam." Lydia's strangely altered eyes dwelt on her husband's face with so fixed a stare, that he felt the low shudder which had once before crept along his nerves.

"I shall, in all probability, take another way," he answered, a threat half revealing itself in his tones. "As just said, I will not have my hard accumulations scattered to the wind. Justice to myself and children demand restriction. It seems that you are bent on carrying things with a high hand. Nearly a hundred dollars spent in three days, and not a word of explanation. No wonder even your children say, that you waste and scatter faster than ten men can earn."

Mrs. Guy started as if stung by a serpent, a sudden paleness overspreading her face.

"My children, Adam?" she said, huskily, and in a voice painful with surprise.

"Yes, your children," returned her husband, with an air of cruel triumph.

"Who said it? What child? When?" There was a trembling earnestness about Mrs. Guy, now.

"I heard it with my own ears; that is sufficient. And when things come to the pass that a woman's children remark upon her wasteful use of money, it is about time for the husband to interfere and save himself from ruin — as I shall do."

This was too hard a blow for Mrs. Guy. She arose, without answering, and left the room. In a few minutes she returned, and handing her husband a small pocket-book, said, in a mild, yet firm voice —

"You will find twenty dollars in that pocket-book, Adam, the remainder of what you gave me day before yesterday. The due bills were in settlement of standing accounts. In the future, you must do all the buying. I shall waste no more of your hard accumulations. What you bring into the house, I will dispense; but not a dollar shall again pass through my fingers. There is such a thing as going too far; and you have stepped over the line."

"Don't play the fool, Lydia," said Guy, impatiently, tossing back the pocket-book, which fell upon the floor. "I've had enough of your silly airs. You're trifling with the wrong man."

"There's no trifling, Adam, as you will find." Lydia was calm, but resolute of manner. "When my children are brought up as false witnesses against me, it is time that I withdraw from a position that has never been satisfactorily administered — and I do now withdraw."

And leaving her husband, Mrs. Guy went to her own room. She had been there only a little while, when her cook tapped at the door.

"There's no coffee in the house, ma'am," said cook, on being admitted; "nor any eggs, nor lard; and I don't think we've sugar enough for breakfast. Shall I run round to the store?"

"No, Margaret. See Mr. Guy, and tell him what is wanted. He will attend to these matters hereafter."

The cook stood in unconcealed wonder, gazing at Mrs. Guy.

"Did you understand me, Margaret?"

"Yes, ma'am. I'm to go to Mr. Guy."

"That is what I said. If anything is wanted in the house, go to him."

The cook lingered for a little while, and then went slowly down stairs. After conning over the matter for some time, and wondering what it could mean, she ventured into the presence of Mr. Guy, who sat in the dining-room, pondering in moody perplexity over this new aspect of affairs. The inflexible persistence of character, united with something in her looks and manner that made him feel uncomfortable, which Mrs. Guy had shown of late, admonished him that trouble was at hand. Margaret entered, and stood before the master of the house.

"Mrs. Guy is up stairs," said he, gruffly.

"It's you that I want to see, sir." Margaret spoke in doubt and hesitation.

"Well, say on."

"There's no coffee, nor eggs, nor lard, sir, in the house — and the sugar's out."

Guy swept around in his chair — he had merely looked at Margaret over his shoulder — and confronted her with a look of half angry surprise.

"Mrs. Guy bid me tell you, sir!" stammered the cook.

"Mrs Guy."

"Yes, sir. I told her about it, and she bid me come to you."

"To me! Aren't you mistaken?"

"Oh no, indeed, sir! She said that when anything was wanted in the house, I must come to you."

"When did she say that, Margaret?"

"Just this minute, sir. I told her what we wanted, and she sent me to you.

"For money to buy them?" said Guy.

"No, sir. She didn't say anything about money. She just told me to come to you."

"Will a dollar get what you want?" asked the perplexed man, diving into his pocket.

"Yes sir," replied Margaret.

Guy handed the cook a dollar, and then went striding, in high feeling, up stairs, to demand of his wife what she meant by all this.

"Nothing more nor less," was her cold answer, "than what I have already declared. You are a hard man for a woman to come in contact with, Adam Guy — a hard, selfish, iron-hearted man! For years I have been wounded and bruised in the contact. Now, I retire from the strife. Flesh has nothing to gain in reacting upon iron. It must, sooner or later, become paralyzed. If gold is your idol, worship on — I shall be no

9

priestess to keep the fires burning on your unhallowed altars."

It was all in vain that Adam Guy stormed, threatened, remonstrated — even persuaded. Lydia had retired from the strife. Folding her arms passively, she sat down, in dreamy introversion of state — taking no care or responsibility in her household, and even becoming strangely indifferent towards, and neglectful of her children. The whole care of the household devolved on her husband, who had to order and superintend, as best he could, in every department. In doing this, however, he had an intelligent auxiliary in Adam, his oldest son, now in his twelfth year — a boy who inherited from his father a strong love of money, with the instinct of hoarding. Guy could trust Adam. So, to this boy was delegated certain functions in the household. He and his father held a conference every evening, and Adam rendered accounts of expenditure in the various departments over which he had control. He, also, in the capacity of spy, kept his father informed of everything that went on during the hours he was at home from school; and often, through the influence of a morbidly excited imagination, of things that had no existence in time and space. Particularly was Adam sharp-eyed in regard to the conduct of his mother; stimulated thereto by the eagerness with which his father listened to every word that threw shadow, blame, or doubt upon her.

So entire a change in the order of life, could not but prove hurtful to a mind already pushed from its even balance. Mrs. Guy's thought and care in her household, under all the painful obstructions that were in her way, were far better for mental health than this dead level, half forced, half morbid indifference. If, in strife with her husband, the powers of an outraged and starved mind were beginning to show signs of failure, the abandonment of that strife, and the giving up of all interest in external things, was to risk the most fatal consequences. Lydia was not in a condition to have the mental strain removed. Safety was in life and action, even though every heart-stroke lifted itself in pain.

CHAPTER XX.

CONCILIATION and adaptation were not the means by which Adam Guy sought to gain any of his ends. Avarice is cruel and pitiless, and guards its treasure in the spirit of a tiger with its whelps. It feels that every approaching footsteps heralds an enemy, and crouches on the alert, always, ready for assault or defence. No matter how weak, or harmless, or innocent the intruder, the talon is surely bared to receive him. It cannot think unselfishly out of itself — has no kindness, no mercy, no generous consideration. All mankind is its enemy. There is no scruple in avarice — only fear of consequences withholds. Whatever stands in the way of its ends, or obstructs as to the means, must be removed if within the bounds of a safe possibility. It tramples on hearts as if they were stones in the street, and is as unmoved by tears, as by the falling of a summer rain.

Such is avarice, and such was Adam Guy. The state of his wife's mind annoyed him, for it was an obstruction. But, it was never once suggested, that this mind was falling into disease requiring the most skillful treatment. Her strange conduct, instead of awakening concern for her reason, irritated him. He was angry towards her, not tender and pitiful. Thus, his treatment still hurt and alienated the unhappy woman. The sentence, " Putting on airs," fitly expressed Adam Guy's appreciation of his wife's conduct. He saw no deeper than that. Avarice made him blind as to any true perception of another's state — more particularly if that state was the result of his action upon the individual.

This sudden giving up of care and responsibility by Mrs. Guy, acted, as we might infer, very unfavorably on herself and family. She fell into a listless, dreamy, wretched state of mind; sometimes weeping in her room for hours; sometimes lying in bed, refusing to answer any questions, or taking food, for whole days; and sometimes wandering about the house, seemingly bent on accomplishing something, and yet doing really nothing. Left almost entirely to the servants, the children did pretty much as they pleased, and soon set all of their mother's occasional feebly exerted authority at defiance. Adam, the oldest boy, acting under instructions of his father, came daily more and more into the office of administrator in household affairs. He received a certain sum of money regularly, and kept an account of expenses, which was nightly examined by his father, and the cash on hand ascertained, to see if it agreed with the balance shown in the accounts. All this was far more satisfactory to Mr. Guy than the previous " loose way of managing things," as he called his wife's mode of disbursing money.

Mrs. Guy, who never set that value upon money which it possessed in the eyes of her husband, had been in the habit of giving pennies and small silver now and then to the children. Adam hoarded, while John spent everything that came into his hands — spent it all for himself. Adam was a selfish miser, and John a selfish spendthrift. The new order of things naturally tended to bring in among the children new causes of strife. Adam, instead of their mother, had the home distribution of money, and in him they found no generous friend. Not a single penny went to them from the closely drawn purse, while many a piece of silver, falsely charged out in the book of expenses, found its way into Adam's money-box. Complaints to their father met with no encouragement. His answer was, that they had enough to eat and drink, and stood in no need of money to spend. Spending was a bad habit, and never should be encouraged by him. Adam took sides with his father against the children, and so they learned to look upon him as an enemy, and to hate him as such.

John, next in years to Adam, was as strong-willed, and as dishonorable at heart as his brother. This sudden cutting off of supplies was a thing to which he was not disposed to submit. He had a mania for spending as decided as Adam's mania for saving, and the means of its gratification must be attained. Up to this time he had enjoyed, through his mother, legitimate means. These being cut of, his thought turned itself in another direction. Adam had a purse, always well supplied with money — the family purse; and John reasoned, that he had rights in that family purse not alienated by any transfer of possession. So, he determined to help himself, at the first opportunity. But no daytime opportunities came. Adam guarded his trust with unwearied fidelity. Money was too precious a thing, in his eyes, to be left a moment unwatched.

John soon saw that only one chance was left. He must finger the purse while Adam slept. So, he kept himself awake one night, until his brother's hard breathing satisfied him that he was in the world of forgetfulness. Then he crept out of bed, and taking the purse from Adam's pocket, abstracted half a dollar, which he placed in one of his pockets.

In making up his accounts on the next day, previously to submitting them to his father, Adam discovered the deficit, and was greatly exercised in mind thereat. The cause was not for a moment suspected. After trying in vain to remember some unrecorded expenditure, he went boldly past the difficulty. Whenever he yielded to temptation, and dropped a coin into his private money-box, the account was made to agree with the balance of money on hand by an entry of some imaginary purchase of sugar, coffee, eggs, or potatoes. This safe method of adjustment came in, naturally, on the present occasion. "Apples" bore one half of John's sin, and "eggs" the other, and the boy went free of all suspicion.

John had a friend in the neighborhood, with whom he passed a great deal of the time not spent in school; and the two lads managed to devour as much cake, candy, and fruit, as the stolen half dollar would buy. Had the money come fairly into the possession of John, he would have shared with nobody. As it was, he felt like transferring a measure of responsibility. Not that he reasoned on the subject — only a blind instinct of safety influenced him, which was as likely to lead into the way of discovery as concealment.

Night found John's pocket empty. The half dollar had melted, under his own and his companion's greedy appetites, like snow in the sunshine. The means of replenishing that empty pocket were again at hand. Not a word in regard to the first abstraction, had been said by Adam, and it was the natural conclusion of John's mind, that it had not been discovered. So, he resolved to take a second step in this guilty direction. After they were in bed, he kept himself awake as on the night before; but Adam seemed as little inclined to sleep as himself. In fact, the loss of that half dollar was troubling him. He could not make it out. A dozen times had he gone over, in his mind, the expenditures of the day, but the missing sum could not be acounted for.

"Adam," said John, after lying still for half an hour, listening in vain for the deep breathing by which he had made him self satisfied of Adam's state of oblivion on the night before.

Adam heard, but, from sheer perverseness refused to answer.

"Adam," John spoke again.

But no motion or sound came from his brother.

"Adam." This time John pushed him, gently. But Adam lay as still as a log, though with every sense on the alert. Why was John lying awake so long? — and why did he speak to him in that hushed way? The very tone of his brother set his thought to questioning, and as the half dollar was pressing on his mind, a suspicion flashed through it. Instead of answering, he mumbled a few words incoherently, like one disturbed in profound sleep, and then commenced breathing in a heavy way.

John, deceived by this, waited a few minutes, and then got quietly out of bed. The room was dark, but light enough came in from the stars for Adam's cat-like eyes to see every movement of his brother. It was impossible for him to wait until the purse, in which he carried the household funds, was opened. Enough, that the hand of John was in the pocket where the treasure lay. Out upon him he sprang, exclaiming —

"So I've caught, you, Mr. John! Aha!"

John was, for an instant, in dismay. The trousers he had taken from a chair, fell to the floor, the purse still in its place. But he rallied himself, as he threw Adam off, replying with affected anger and scorn,

"Aha, what?"

"Thief! Robber! You stole half-a-dollar last night!"

"It's a lie!" answered John, boldly.

"I'll tell father all about this in the morning, Mister; and he'll make you smart! I wouldn't be a sneaking thief!"

"If you say thief again, I'll knock you over!"

"Thief"— Adam hissed back into his brother's face, who struck him in blind passion. Both lads now forgot every thing in the angry strife that followed. Adam was oldest and strongest; but about John, when excited, there was a wild desperation, that, on the first outbreak, bore down all resistance. The blow with which his blow was answered, aroused him to fury, and flinging himself upon Adam he drove him backwards upon a chair over which he fell with a loud noise, and a louder outcry. This brought Mr. Guy, not yet in bed, to the scene of trouble.

"What's all this about?" he demanded in angry tones, as he pushed open the chamber door, and let the light from a passage lamp stream inward.

"He called me a thief," answered John, getting in the first defence.

"And so you are!" replied Adam, boldly. "You stole a half dollar out of my pocket last night, and"—

"It's a lie!" fiercely retorted John.

"It's the truth," persisted Adam, "and I caught you in the very act of robbing my pocket again to-night."

"Is that so?" demanded Mr. Guy, a cruel sternness in his voice.

"Yes, sir. It is so."

"It's a lie!"

"Silence, sir!" Mr. Guy raised his hand.

"Indeed, father, it isn't true." John's voice changed to one of piteous denial.

"Adam, I want the truth of this matter," said Mr. Guy, turning to his oldest son. "You say that John took half-a-dollar out of your pocket last night."

"Yes, sir."

"No, sir. I didn't."

"Silence, I say! And you caught him at your pocket again to-night?"

"Yes, sir. I missed half-a-dollar this morning; and

to-night I kept awake for a good while after I went to bed. I thought John was asleep, for he breathed as if he was, when he called me. I didn't answer. Then he called me again, and pushed me. But I kept still, and pretended I was asleep. After awhile, he crept softly out of bed, and I watched him go to my trousers and begin looking for the pocket. At this I darted out on him and he struck me in the face."

Mr. Guy waited to hear no more. Adam's story was fully credited, John tried to explain that he had a cold, and was after his pocket handkerchief; but his father caught him with a vice-like grip and gave him a terrible flogging.

"You stole the money yourself, and lied me into a beating," said John, sobbing from pain, as he crept back into bed after his father had left the room. "But I'll fix you for it, see if I don't!"

"Fix away! Nobody cares for you!" retorted the hard-hearted Adam. "If I'd been father, I'd have given you twice as much."

Thus they snarled at each other like two wild animals until sleep overcame them, and both sunk away into that oblivion of outward things that comes as a blessing to old and young — to the evil and the good — to the conscience-clear and the innocent.

CHAPTER XXI.

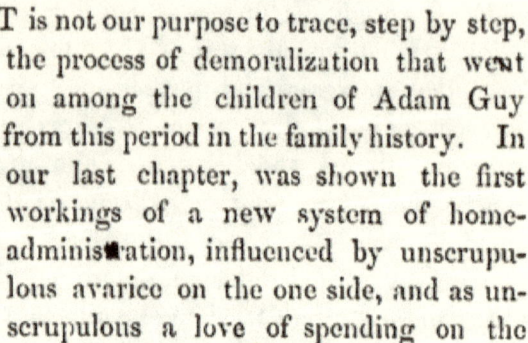IT is not our purpose to trace, step by step, the process of demoralization that went on among the children of Adam Guy from this period in the family history. In our last chapter, was shown the first workings of a new system of home-administration, influenced by unscrupulous avarice on the one side, and as unscrupulous a love of spending on the other — two powers in never sleeping opposition to each other. Young Adam, elevated to a seat of executive domination, matured in the one direction so constantly stimulated with unnatural rapidity, while all the children regarded him with dislike, as a usurping and oppressive tyrant, and set themselves against him on every occasion.

John's love of spending was in no degree lessened by the new difficulties which had come in his way. Adam guarded the family purse with a fidelity that only permitted occasional abstractions therefrom. The amount thus obtained, fell so far below what our young spendthrift desired, that he set his wits to work in a new direction. Having begun a dishonest course, there was no question in his mind as to the right or wrong of any scheme that suggested itself. Passing a pawnbroker's on the way from school one day, he asked Adam what kind of a business the man who kept the shop carried on. Adam explained its nature, but, ere he was done, the mind of John was grasping the general idea involved, and turning over its suggestions of ways and means for supplying his empty pockets. On that very afternoon, he took a breastpin from his mother's drawer, and pawned it for a dollar and a half. The ticket received therefor was destroyed.

Thus began a course of evil on the part of this unguided boy, destined to involve his manhood in ruin. No very long time elapsed before discovery and its penalties followed his criminal conduct. But, this only obstructed his purposes, and set his wits to work in other directions. Money still continued to find its way into his pocket, and by means which, when brought to light, as was often the case, subjected him to punishment and disgrace. Under this he hardened, so as to endure and grow defiant, but changed not in the smallest degree.

Mrs. Guy, after dropping the reins of family government, never attempted to gather them up again. Now and then, as the appearance of sudden danger startled her dull mind, she would grapple at the reins as we sometimes see a scared woman in a carriage, and put all in greater danger. But, for the most part, she moved in the household as one who had in it only a partial interest, and no controlling power. Towards her husband she maintained a cold reserve, never intermitted under any circumstances. If he attempted arrogant enforcement of his will towards her, she defied him with an outflashing anger that blinded and half scared him, like a gleam of fierce lightning. Money she would not touch. He laid it in her way over and over again; but it remained without appropriation.

There was another change in Mrs. Guy, more remarkable than the rest, because its source was not apparent. It consisted in an entire alienation towards Mrs. Hofland. The renewal of their personal intercourse had seemed to promise much for Lydia. The healthy, cheerful, clear-seeing mind of Lena, was just the one she needed as a companion-mind. But, after receiving a return visit from Lena, she did not call again at the peaceful little home to which she had gone in yearning pity for her friend's supposed misfortune; nor, at Lena's subsequent visits, would see her. Alone — all alone as to companionship or sympathy, she sat down in her gilded prison, and denied herself to every visitor.

Startled, at length, by some more palpable evidence of insanity, the thought of an asylum entered the mind of Adam Guy. Once there, that thought became a permanent guest. He looked at it, dwelt upon it, turned it over and over, and finally accepted the suggestion as pointing to the easiest and most effectual way of getting an obstruction removed from his path. It is said that "the wish is father to the thought;" and the proposition was no doubt true in the present case. Mrs. Guy was a source of constant trouble and annoyance to her husband, and it was but natural that he should feel a desire to get freed from this unpleasant accompaniment of his daily life. If death had glided in to solve the difficulty, he would have accepted the fact, and bowed in cheerful submission. But, death was not ready to attend Adam Guy as a minister, and quietly remove from his house the woman who had ceased to be anything to him but a thorn and a hindrance.

An asylum was the only resource. Adam was not wicked enough or desperate enough, to be in any direct way an accessory to death; but in the matter of his wife's removal to an insane hospital, he could act with a clear conscience — we use the word conscience in a very low and natural sense — and he soon set himself deliberately at work to compass this result. He had no trouble, after very exaggerated statements of his wife's case, in getting a certificate from his family physician, declaring her out of her right mind, and a fit subject for treatment. But the prime difficulty lay in her removal. If she had been in the habit of riding out, all would have been easy enough. But, Mrs. Guy never passed the threshold of her own door. The idea of force came now and then into her husband's thought, as he grew more and more impatient to get her out of his way; but, he feared to attempt this, lest there should be violent resistance, and exposure in the neighborhood.

Days, weeks, and even months passed, after Guy received the doctor's certificate declaring his wife a fit subject for treatment in an asylum, and still no opportunity for removal was presented. He was growing desperate under this long delay. In his heart, Lydia was repudiated; she had become hateful in his eyes; every motion was an offence; she was a skeleton in his house — a death's head at his table. Each act towards her was a studied wrong, prompted by an ill-repressed anger. He sought to drive her from incipient derangement, into strongly defined insanity, which would make forcible removal from his house a necessity; but a dead level of indifference and contempt was her protection.

One evening, it was nearly a year after the unhappy change in Mrs. Guy's state of mind, to which we have referred, Guy came home and found his wife suffering terribly from neuralgic pains in the head and face. Her agony was so great, that she walked the floor incessantly, tears, wrung out by intense bodily anguish, flowing over her cheeks. Guy affected pity, though he felt none, and with a show of kindness, that was but a cover for a suddenly suggested plan of carrying out his long-cherished design, started off to see their physician, and get, as he said, something to relieve her pain. In half an hour he returned, with a small vial, from which he gave his wife a few drops. She knew it to be morphine, but, suspecting no wrong, and being almost wild with pain, took the potion in hope of relief. In ten minutes, a second dose was administered, and in ten minutes afterwards, a third, which soon after locked the already obscured senses in a sleep that imaged death.

White, as if the spirit had departed, lay the form of his wife before the face of Adam Guy, as he stood alone with her in the chamber where she was now fully in his power. Was there pity in his heart? Did the old-time feeling come back upon him? Did he soften in this sad presence? No — no — no! Avarice never softens. Neither fire nor water subdues its triple induration.

What next? Adam could proceed no further without help. He stood and thought — moved about the chamber irresolutely — stood and thought again.

For the past six months, the family had been in charge of a housekeeper, towards whom Mrs. Guy maintained an unwavering hostility. But the woman kept her ground, without showing any ill-will or disturbance. She was cold in manner, orderly in her habits, and, while seeming not to assume authority in the family, steadily reached out for the governmental-rein, and held it with a quiet strength that overcame resistance not roused to passion by a display of power. Her name was Mrs. Harte. She was a widow, age about thirty-five; had been a widow for over seven years. During her residence in Mr. Guy's family, there had been no familiarity between her and the master of the house, who held himself aloof in cold dignity, only conferring with her in matters strictly limited to her administration. The boy Adam, continued to be the disbursing agent. Mr. Guy was not going to trust his money in the hands of any one from whom he could not exact the strictest account. Whatever opinion Mrs. Harte might hold in respect to Mrs. Guy's state of mind, and the causes leading thereto, was the result of observation alone. Not a word had passed between her and Mr. Guy on the subject.

After nearly five minutes of debate, Mr. Guy took hold of the bell-cord, and gave it a quick jerk. He stood close to the door, and when a servant came, opened it a little way, and said —

"Ask Mrs. Harte to come here."

A small, compactly-built woman, with a gliding step, entered a few moments afterwards. There was a clearly cut outline in every feature of her intelligent face, showing decision and firmness of character. Her complexion and hair were light, and her eyes a pale, cold, almost leaden blue.

"What is the matter, sir?" she asked, in a tone of surprise, as she glanced towards the bed, on which lay the unconscious Mrs. Guy.

"The effect of morphine," said Mr. Guy, calmly.

"Morphine!" An expression of doubt and concern came into the face of Mrs. Harte.

"She has been suffering terribly, with that *tic* in her face."

"Yes, sir — I know."

"I went round and saw the doctor, and he sent her this," taking up the vial from which he had given the medicine. "It has produced a condition of physical insensibility, as he desired. Poor thing!"

There was a tone of pity in the voice of Mr. Guy.

Mrs. Harte went to the bed, and stooping over the insensible woman examined her carefully.

"How much did she take?"

"Only three small doses. Her system is not used to narcotics, and has yielded quickly."

After looking at Mrs. Guy for a few moments, Mrs. Harte turned and fixed her cold, blue eyes on the face of Mr. Guy. He saw inquiry in them.

"What is to be done?" There was an invitation to confidence in her voice clearly apprehended by Mr. Guy.

"Sit down, Mrs. Harte." The offered chair was accepted. Mr. Guy went back to the chamber door and turned the key. Then he drew another chair in front of the housekeeper and sat down. She looked calmly expectant.

"May I claim your confidence, Mrs. Harte?"

"Yes, sir. I am a discreet woman."

Guy fixed his eyes intently upon her. She did not betray a sign of feeling, nor turn away from his her cold gaze for a moment.

"You have not failed to observe my wife's unhappy state of mind?"

"It is too apparent, sir, to every one," answered Mrs. Harte.

"Nor, that it is on the increase?"

"Unfortunately, its increase is undoubted," said Mrs. Harte.

Mr. Guy now took from his pocket a folded paper.

"Read that." And he handed it to Mrs. Harte.

"The Doctor's certificate, I see."

"Yes."

"Pronouncing her insane, and in need of treatment in some Asylum."

"Exactly. And you will see, by the date, that I have had it in my possession for several months."

"So I perceive."

"Forbearing all that time, and hoping all that time for a change in her condition, that would render needless a last resort."

Mrs. Harte sighed very naturally, adding to her sigh the words — "Poor lady!" and then shook her head in a hopeless kind of way.

"This is a fair opportunity to have her removed," said Mr. Guy, comprehending the woman's state of acquiescence. "She is entirely unconscious, and will, in all probability, remain so for hours. Can I secure your coöperation?"

"If all is right," answered the woman.

"You hold the guaranty in your hand, Mrs Harte. As to the insanity, your own observation makes that clear."

"That is clear enough, poor lady!" said Mrs. Harte.

"And the authority for her removal is explicit."

"So I perceive."

Then let us act without hesitation or delay, both for her sake and that of her children, over whom her influence is of a very unhappy character. My purpose is to remove her to-night, and have her safely cared for before consciousness returns. What do you think of it?"

Mrs. Harte reflected for a few moments, and then replied,

"I do not see, sir, that a better opportunity is likely soon to occur. You are certain that she has not taken too much of the morphine?"

And Mrs. Harte gave Guy a searching look.

"Too much! No! I kept to the Doctor's prescription within a drop."

"Because," added Mrs. Harte, in her calm, clear voice, "I do not wish to get myself into any trouble."

"There can be no trouble to any one in the case," said Mr. Guy.

The woman's eyes did not fall away from his face, but

dwelt on it for several moments, and with an expression that Guy felt as a kind of power over him. He had not, since this woman came into the family held with her, until now, any familiar or confidential intercourse. Wrapped in his own separate thoughts and interests, he had moved about his house, only considering Mrs. Harte as a useful appendage in her place, and of no more account to him than any other bit of domestic machinery. But, now, the relation had changed, and he felt it — felt it with an inward sense of reluctance and repulsion.

"If there is entire safety, sir." How even and penetrating her voice! — How steadily her cold, searching eyes rest upon his face!

"You understand the case as well as I do, madam." There was an apparent drawing back from the woman, on the part of Mr. Guy, which was perceived on the instant, and produced a change in her manner.

"I believe so, sir," her tone was softer and more acquiescent; "and if I can serve you and the poor lady in anything, I stand ready to act. She will be a great deal better off in a well-managed Asylum. Where do you think of placing her?"

"Among the Sisters at Mount Hope."

"Ah?"

"Yes. I have already consulted them on the subject, and shown the Doctor's certificate. They are prepared, at any time, to receive her."

"She could not be sent to a better place," said Mrs. Harte.

"I am sure not. And my instructions will be that she receive the kindest and most humane attentions.

But, time passes, and we must act promptly, if we act at all."

"True, sir." And Mrs. Harte arose.

"Have the children all in bed before her removal," said Mr. Guy.

"The three youngest are asleep; but you will have to see after Adam, John, and Lydia. If I were to say 'go to bed,' they would sit up half the night in rebuke of my assumed authority."

"Very well. I'll settle that. Remain here, and get all things ready for her removal. After the boys and Lydia are in bed, I will go for a carriage. You must be prepared to accompany us. Lock the door when I go out, so that no one in the house may intrude and obtain a knowledge of what is going on."

Mr. Guy then withdrew, and Mrs. Harte turned the key.

CHAPTER XXII.

HE woman's aspect changed instantly, when alone. The cold eyes and face flashed and gleamed; the placid manner became disturbed; a look of satisfaction — almost of triumph — flitted over her countenance. Quickly she proceeded to the work of changing the garments of Mrs. Guy, and getting all things in readiness for removal. Now and then, she would stop and consider the pale, death-like face before her, — not in pity; not in fear; not in hate; but with a look of searching inquiry, in which doubt and desire blended. There was a covert eagerness in her manner, seen in the unusual celerity with which she hovered about the bed, and the occasional unsteadiness of her hands as they moved over the person of Mrs. Guy. Her part of the work was done, long before Guy was ready, or the carriage at the door.

"Has she stirred yet?" was the whispered question of Mr. Guy, as he came into the chamber, when all was prepared.

"No, sir."

Guy crossed to the bed, where his wife lay, and stood regarding her for a few moments. An image of death, not life, was before him. His heart gave a strong throb, and he turned a face of alarm upon Mrs. Harte. They looked at each other in silence for some moments.

"Her heart beats," replied the woman, who understood him.

Guy took his wife's small, wasted hand in his, and placed his fingers on the wrist.

"I don't find any pulse," said he, turning pale. His voice was disturbed.

"Lay your hand over her heart."

Guy obeyed the suggestion.

"Don't you feel it beat?"

"No!"

"What!" And Mrs. Harte thrust her hand in, pushing that of Mr. Guy aside. Suspended breaths marked the intense interest of both. "It beats, sir! There! Put your hand there!" She spoke in a whisper, quickly.

"Yes — yes. I feel it! But how very low and faint," said Guy, as he withdrew his hand and stood up, in doubt and irresolution. Then he laid his fingers again over the artery on her wrist. Not the feeblest thread of motion touched the alert sense of feeling.

"You gave her too much, I fear," said Mrs. Harte, letting her pale, blue eyes rest firmly upon him.

The face of Mr. Guy turned still whiter.

"In that case" — and the woman made a step backwards, pausing with the sentence half finished.

"How in that case?" Guy felt himself already in her power.

"She may not rally," said the woman.

"The Doctor is responsible. I only followed his prescription."

"How often did she take the medicine?"

"Three times."

"How much did you give her each time?"

"Only a few drops."

Mrs. Harte crossed the room to where the vial of morphine still remained on the mantel piece, and taking it in her hand, held it up to the light. It was one-third empty.

"Was it full when you received it?" asked Mrs. Harte.

"No." But the manner of his answer betrayed the truth to Mrs. Harte.

"There has been an over dose," she said, confidently. "I'm afraid you mistook the amount of vital power in her system."

"If there has been a mistake, it lies at the Doctor's door, not mine," answered Guy, in too apparent alarm.

"No dose is marked on the label." The woman's eyes turned from the vial, and again dwelt, searchingly, on Guy's face. He quailed a little, and she saw it.

"I think," said Mrs. Harte, speaking with deliberation, "that I understand the case, which has assumed a very serious aspect. You did not see the Doctor at all."

Guy started, frowned, and was about to repel the assertion, when the housekeeper lifted her hand, saying, with perfect coolness —

"A moment, sir, if you please. If the Doctor had ordered morphine, the dose would have been indicated on the prescription, and the druggist would not have failed to mark it on the label. The truth is, you procured the drug without consultation, and administered it in too strong doses. Am I not right? You had better trust me fully, if you want my aid. I must see where I stand, and the actual peril in my way; before I take a step forward. You have put yourself in a very dangerous position, sir."

A tremor seized the nerves of Mr. Guy. The feeble denial attempted broke down ere half of it was spoken. He could not go on, with the cold, searching eyes of a woman, whose character he was only now beginning to understand, resting on him like a spell.

"You meant no harm — only good," continued Mrs. Harte. "That I understand clearly. But intentions are out of the pale of consideration now. Actions and effects will only be regarded. In order to get Mrs. Guy into a condition for removal to an Asylum, you gave her morphine."

"The morphine was taken to relieve pain," said Guy.

"Ostensibly. That was the plea. The real object was to accomplish something beyond. You gave her too much, and her life is in peril. Now, do you want my assistance or not? Heaven knows, if I consulted my own feelings, I would pass, instantly, from under your roof. Say that you do not care for my aid in the matter, and a heavy burden will be removed. But, being here, I will not shrink from duty, if called on for help."

"What shall be done?" asked Guy, showing symptoms of a helpless bewilderment of mind. "It will hardly be safe to remove her to Mount Hope, seeing that she is sinking so rapidly."

"No. She might be dead before we reached there. The Doctor had better be called in."

"Is that necessary?" asked Guy.

"Yes. It will not do to make any mystery with him. Let him understand the naked facts in the case. The maddening neuralgia, and the doses of morphia. Say nothing of course, about any object beyond."

"Hadn't I better say, that she took the doses herself?" asked Guy, actually trembling in the face of peril.

"No. All attempts at concealment will involve greater danger. Let the Doctor clearly understand the case, and he will exonerate you from blame, and give the medical certificate required for burial in case she does not live."

Mrs. Harte took up the vial of morphine again.

"Are you certain it was not full?" she asked.

"Positive," answered Guy.

"The druggist *may* be referred to." Mrs. Harte looked at him meaningly, and he understood her.

A pause ensued, in which each regarded the other.

"There is alcohol in the house," said Mrs. Harte, breaking the silence, "and it will be prudent, I think, to add a couple of spoonfuls to this vial. Go, or send for the Doctor. I will see that the vial is replenished."

"Then, it is understood that the removal is abandoned for the present?"

"Of course. That cannot be thought of for a moment. The Doctor must be had, and speedily. Don't lose another instant, sir. Everything depends on promptitude now."

Thus enjoined, Mr. Guy went hastily out, and jumping into the carriage which was in waiting at the door to remove his wife, ordered the driver to go with all speed to the residence of his family physician. In twenty minutes afterwards, the Doctor and Mr. Guy entered the chamber where the unconscious woman was lying.

"It is too late," said the Doctor, after sitting for a little while at the bed-side. "No human skill can save her."

"In heaven's name, no, Doctor! Don't say that!" And Guy exhibited what seemed uncontrollable anguish. "You must save her!"

But, the Doctor shook his head soberly, and then asked —

"How much did she take?"

"Only a few drops, sir, as you can see," replied Mrs. Harte, producing the vial. The Doctor held the vial towards the light, and examined it for a moment; then handed it back with the remark —

"The symptoms indicate a much larger dose."

"Was there no heart disease?" asked the housekeeper.

The Doctor turned and looked at her sharply for a moment; but her cold eyes did not shrink nor waver.

"You gave an emetic, I understand?" He looked from the woman to Mr. Guy.

"Oh yes, sir, immediately on finding that she had apparently taken too much."

"But no action of the stomach followed?"

"None whatever."

"What else was done?" The Doctor referred to Mrs. Harte.

"Nothing, sir. We were so alarmed and confused — we did not know what to do. The effect of a few drops was so extraordinary. I've never seen anything like it in my life. There must have been some organic trouble."

An attempt was now made by the Doctor, to give an emetic, but without effect. Death was too near and too certain. In less than half an hour, the curtain fell over this tragedy of life, and a weary head and an aching heart were forever at rest.

CHAPTER XXIII.

THE Doctor was not satisfied in regard to the death of Mrs. Guy; yet there was no evidence of foul play, and he pushed aside the doubtful questions that kept intruding themselves. His certificate, in the usual form, made all plain for the burial, and left no room for suspicion to take any dangerous aspect in the public mind. People talked as they always will talk when there is a shadow of mystery, and many idle stories were whispered around; but, the real truth did not transpire.

For a time, Adam Guy felt a sense of freedom. An obstruction, which had hindered and annoyed him for years, was removed. He was master of the position once more. His will could go forth unquestioned at home as well as abroad. But, soon, he became aware of a presence in the house that touched his freedom more vitally than it had ever been touched. That presence, though neither demonstrative, nor obtrusive, became more and more palpable as a fixed fact. The quiet, self-possessed, cold, orderly housekeeper, sat at his table every day, silent for the most part; and moved through his home with a power of subjugation that rarely provoked resistance, yet steadily intruded itself, gathering up the reins of government, and preparing to hold its place with a strong hand when the time for throwing off the mask came.

Adam Guy was in the power of this woman; and he began to have an unpleasant consciousness that she meant to use the power in some way, to her own advantage. She did not seem to have a will of her own in the family, and yet, Guy saw new movements going on in the order of things which no hand but hers directed. Occasionally, she would suggest a change in presence of the children, which brought from them opposition or remonstrance. She did not argue the case, nor show special interest in the mattter — but, it usually happened, that Mr. Guy came over to her side; not because he saw clearly any special value in the changes proposed, but from an indefinable impression of power that he did not feel willing to oppose. Rebellion, however, soon began to lift its head among the children, to whom, during their mother's life-time, the housekeeper had yielded with a skill or passiveness that rarely provoked opposition. Not until she had a firm grasp upon the rein, did Mrs. Harte begin to draw upon it so steadily as to make the pressure felt. The father was in her toils, and she would now reduce the children to obedience. But the task, in this case, was more difficult. She had no command of their fears. There was not in their hearts a fatal secret to which her finger ever pointed in warning. When they looked into her face, they did not see "Beware!" fluttering on her lips.

Adam had always treated her in a half insolent way

as if she were inferior and of little account; and usually took pains to give things a direction, if possible, adverse to her wishes. This he was able to do, in consequence of his position as purse-bearer and account-keeper for the family. From this position, Mrs. Harte meant to have him removed; but, she was in no haste. First of all, she must be well seated. There must be no doubt as to her influence over Mr. Guy.

"How much do you give Adam for housekeeping purposes?" She put the question one evening when they sat alone together. Mr. Guy had no intimate acquaintances, and so went out in the evening but rarely. Mrs. Harte had him, therefore, so much in her power; and in the most unobtrusive and unapparent way, managed to interest some of the lonely hours he spent at home.

"Twenty dollars a week," answered Mr. Guy.

Mrs. Harte looked down at the work in her hands, and remained silent.

"Why do you ask?" After a pause of nearly a minute, Mr. Guy put this question. He had been waiting all that time for the housekeeper; but she did not seem inclined to any further remark.

"Adam manages things very well for a boy — better, probably, than one boy in a dozen could manage. Still, he is only a boy, and cannot be expected to understand the requirements of a household like this. Twenty dollars a week, if spent with judgment, should go farther than he makes it go."

Mr. Guy did not answer. He felt a shade of perplexity coming over his mind.

"It is nothing to me, of course," resumed Mrs. Harte. "I only express a passing thought. Another thing I have observed is this. The position occupied by Adam, puts him in a kind of forced antagonism towards the other children. Between him and John there is a constant feud, growing mainly out of the fact that Adam's will regulates much in the family. He markets to suit his own tastes and whims; and, I observe, takes pains to omit buying the very articles for which John and Lydia manifest a preference. If it goes on, a permanent alienation between him and the other children, as they grow up, will assuredly take place; and, of all things, this should be guarded against as the worst of evils."

"Is that so? Does Adam really annoy his brothers and sisters in the way you intimate?" said Mr. Guy.

"Have they never complained?" asked Mrs. Harte.

"O yes. They're always complaining. But I'm used to that, and pay little heed to what they say."

"They have cause." The housekeeper's voice had a shade more of feeling. After a moment she added,

"One child in a family should not have as much power over the other children as Adam now possesses. He does not know how to rule wisely; and they live daily in a state of half angry rebellion against him. Besides, sir, the boy's mind should be educated towards a man's duties in life, and not towards a woman's. You design him for regular business — for a merchant, like yourself — not a boarding-house keeper. His present office at home is not, therefore, good for him. It will belittle his mind — narrow it down to the smallest things — incapacitate him for the larger sphere in which you look to see him move. But, excuse me, sir, for this freedom of speech. I have been led to say more than I intended."

"There is reason in what you urge," returned Mr. Guy, "and I must think it over. Adam is a little inclined to be overbearing, I know; but, as affairs have been, I could do no better than place things in his hands; and, all circumstances considered, it seems to me that he has managed admirably. Not one boy in ten would have done so well."

"Probably not one in a hundred," answered Mrs. Harte.

And there the conversation dropped. But Mr. Guy understood his housekeeper. She was not satisfied to remain any longer subject to the will and direction of a boy. He must pass out of her way.

Mr. Guy did not act immediately on the suggestion of Mrs. Harte. Too much was involved in this. It included the fact of a new disbursing agent in the household, and that agent the lady herself. Would it be wise to admit her to this place of power? Over, and over, and over again the question was revolved, and yet without decision. In the meantime, Mrs. Harte, from behind the screen of an unimpassioned exterior, watched, eagle-eyed, the progress of things, drawing all the while a little and a little more firmly on the reins of government that were in her hands.

"Adam," said she, one morning, as the boy passed her in the hall. She knew that it was market day.

He stopped, turning his head partly towards her, with an air of indifference.

" Are you going to market ? "

" Yes."

" I wish you would get a pair of chickens for to-day's dinner."

" I shall get corned beef;" was almost insolently answered.

" Oh, very well." And Mrs. Harte turned from him in her calm, quiet way. An observer would have detected no indication of a quicker heart-throb.

At dinner time, Mrs. Harte said, speaking across the table to Mr. Guy —

" I asked Adam to get a pair of chickens for dinner to-day." Her tones were cold and even.

Mr. Guy turned his eyes on Adam, whose face colored a little.

" Why didn't you do as Mrs. Harte desired ? " The boy was not prepared for the sternness with which this question was asked, and stammered out an unsatisfactory reply.

" Don't let it occur again." Mr. Guy spoke in earnest.

No more was said at the time; but the spirit destined to rule in that house had gained a victory, and soon every inmate had an impression of the fact. It took only a week or two from this time for Mrs. Harte to bring Adam to the position of a mere agent of her will in the household administration. He bought as she gave direction, being little more than purse-bearer. There were no contentions between him and Mrs. Harte. If he rebelled, and was insolent, she did not stoop to his level, but, with subtle management, turned his father's iron hand upon him. In six months after the death of Mrs. Guy, the housekeeper's will was supreme in the family where wife and mother had been thrust aside, and held as of no account.

CHAPTER XXIV.

RS. Harte was a woman of superior mind, and some cultivation. Her husband, a lawyer of considerable promise, died just as he was rising to an eminent position at the bar, and she was left without an income. In order to sustain herself, she taught for a few years; but not finding, in this occupation, anything congenial, she gave up her scholars, and accepted the place of housekeeper in Mr. Guy's family. Her position there proved very far from being agreeable, and she was simply waiting for an opportunity to change, when the grave incidents attendant on the death of Mrs. Guy altered her purpose, and she determined to remain. A rich man was in her power, and that was an advantage not to be lightly thrown aside — an advantage which she was just the woman to accept.

Mrs. Harte was as thoroughly selfish in her ruling quality of mind, as Mr. Guy was in his; and more subtle and cruel. He bore down opposition, when it presented itself, with the strong hand of conscious power; but she wrought stealthily, gaining her ends by almost unapparent intrusions, and revealing them only when accomplished. Both loved money; but his love of money had its foundation in avarice, while hers rested on ambition. He desired money for its own sake; she for the sake of power, position, and influence. Pride ruled with her, avarice with him.

While it was not her intention to leave the house of Mr. Guy, she managed to give him the impression that she only waited for an opportunity to make a change. Most cunningly did she bend, at times, the brief passages of conversation which passed between them, so as to touch that memorable death scene, and always with a hint of overshadowing peril that sent a chill of fear to the heart of Adam Guy, and made him more distinctly conscious that he was fatally in her power. She was a kind of terror in his house; yet, on no account would he have that terror removed. A secret lay hidden in her bosom, the revelation of which might, at any time, put even his life in jeopardy. She must, therefore, be conciliated, and kept in friendly contact. There was safety in her friendship; but disaster in her enmity.

And so, as Mrs. Harte gradually assumed a controlling position in Mr. Guy's family, Mr. Guy receded and left the way clear, taking her side in all contests with the older children, and compelling their acquiescence to her rule.

Only a few months were needed, under the new order of things, to make Mr. Guy aware of the fact, that so far as his home-comforts were concerned, he had gained sensibly by the death of his wife. For a year or two, Lydia had been so indifferent towards him, that she neglected many personal attentions; and left him to care for himself. But now, a woman's thought and a woman's hand were becoming more and more apparent in all things of his wardrobe, and in all the home arrangements that touched him personally. The younger children were kept more quiet when he was in the house, and the injunction was constantly falling on his ears, given half aside and in a hushed voice, — " Don't do so; it will annoy your father,"— or words of similar import.

It was in vain that Mrs. Harte strove to conciliate or bend Adam, the oldest boy, to her will — and she was equally unsuccessful with the sharp-tempered, clear-seeing Lydia. Her usurpation of Adam's prerogative in the family, was an offence neither to be forgotten nor forgiven. The boy had tasted of power, and could not relinquish it and accept submission. He was, therefore, a rebel in heart, and, on all suitable opportunities, a rebel in act. But, he put himself in antagonism to one against whom he strove in blindness, losing power in every struggle. Lydia, who, with all her ill-nature and waywardness, had loved her mother, could not bear to have a stranger take her place in the household. She was a girl of quick perception, and saw deeper than any one into the character of Mrs. Harte; or to speak more accurately, had a truer impression of her character and designs. John had a weaker side — he was a sensualist and a spendthrift — and Mrs. Harte, by occasional indulgences, won him over to her will.

Adam and Lydia, who felt, daily, that Mrs. Harte was gaining strength, united their forces in a league against her, and inaugurated an undying warfare. They watched her every movement, interpreting to each other her words and actions to suit themselves, and putting all manner of obstructions in her way. But in every move she thwarted them without a seeming effort, turning, often, their machinations to their own discomfiture, and leading them, by a few well directed questions, to an exposure of themselves to their father. So things went on, the antagonism between Mrs. Harte and Adam and Lydia, gaining strength all the while, but taking on no appearance of anger, dislike, or stern resistance on the part of the former. She was always cold, calm, dignified, and to the eyes of Mr. Guy, in the right.

After the lapse of a year, Mrs. Harte's position was so well assured in her own mind, that she began to act with less caution. Up to this period, she received only the wages of a housekeeper, according to the original contract, — twelve dollars a month. It was time to have some new and better arrangement — to get nearer the full coffers on which her eyes had dwelt with covetous longings. Mr. Guy seemed too well satisfied to let things move on as they were going. She was administering all his affairs with order and economy, and home was more comfortable than it had been for years. Why should he desire a change? For twelve dollars a month, he received a liberal service, and was satisfied. Not so Mrs. Harte. Patient waiting was at an end; for she saw no sign of any new order of things in the family.

It was now that an incident occurred that aroused her to immediate action. One day, while walking in the street, she saw in advance the familiar form of Mr. Guy, in company with a lady. Even her calm pulse leaped.

What did this mean? Who was this lady? Her dress was elegant, and she walked with a self-conscious air. Mrs. Harte checked her pace, and lingered a little way behind them. They appeared to be in familiar conversation, as if well acquainted. For the distance of nearly two squares, Mrs. Harte kept them in close observation; then they stopped at the corner of a street, and after talking a few moments, separated. At parting, Mr. Guy bowed with considerable formality, and with the air of one who evidently sought to make a good impression.

The lady, who turned off from the main street, glanced back, two or three times, before Mr. Guy was out of sight.

"Who is that?" came audibly from the lips of Mrs. Harte, as she stood still, like one who felt a sudden shock. Then pressing forward quickly, she followed the lady, and after passing her, turned at the next corner and stood, as if in doubt. The fair, attractive face of a woman scarcely beyond thirty years, looked almost smilingly, yet a little curiously, into hers. Mrs. Harte was at fault; she did not know the lady. *But she must know her!* Remaining in apparent hesitation until the stranger moved onward a short distance, she then followed slowly at first, but quickening her pace so as not to be thrown too far behind. All at once she lost sight of her. The lady had turned into another street. Mrs. Harte hastened forward with accelerated steps, but when she reached the corner, the object of her pursuit was no where visible. She had entered one of the fine dwellings that stood in the block.

Baffled, and excited with strange alarm, Mrs. Harte retraced her steps, and took her way homeward. The countenance of that lady was full of winning grace. Who was she? What was she? Wife, widow or maiden? Hurriedly these questions chased each other through her mind. Was she rich, as well as attractive? — a widow, as well as beautiful? Ah! what had a poor housekeeper, with few personal charms, to hope for in a rivalry here?

"I must know who she is!" Mrs. Harte said this in a resolute way, forcing back the tremors that were agitating her. And then she grew calm, and self-possessed, and clear-seeing. If this were an obstruction in her way, it must be removed. But, first, to be assured that it was an obstruction. On the next day, on the next, and on the next, Mrs. Harte visited the neighborhood where the lady had disappeared, in order to ascertain, if possible, by seeing her at a window, or going in or coming out, in which house she resided, and thence her name and position. Three times her visit failed of any satisfactory result; but on the fourth day, in passing down the block, she saw the lady descend from one of the houses, enter a carriage, and drive off. The name on the door was noted. It was Leslie.

"Mrs. Leslie!" Her heart bounded. She had often heard of this lady, a widow, holding in her own right, a large fortune. Ah, here was a formidable rival indeed, if rival at all! Rich, elegant, attractive — what had she to offer in opposition to these?

Of late, Mr. Guy, after dressing himself with scrupulous care, went out, occasionally, in the evening. The fact had already awakened a feeling of uneasiness with Mrs. Harte; now, the circumstance presented an alarming aspect, for it was connected in her mind with visits to the rich young widow. On this very evening — the one following the day on which Mrs. Harte discovered the lady's identity — Mr. Guy dressed himself and went away. As he left the house, Mrs. Harte passed to her own room, where she moved about restlessly for some time. Then she sat down, with deep lines on her ordinarily smooth brow, and a tight pressure on her lips, that were firmly drawn against her teeth. Her hands lay clenched upon her lap.

"Never! Never! Never!" The words came in a deep whisper, while a gleam of passion quivered over her face. "I will not be pushed aside by any one!"

"Rising, she went to a drawer, and unlocking it, took out a vial — the same from which Mr. Guy had administered the morphine to his wife — and held it to the light. It was nearly full. The reader will remember, that to hide the fatal secret of an overdose, she had added alcohol, and so deceived the physician. But, now she poured from the vial a portion equal to that added.

"This is my argument," she said, as she recorked the vial, and held it again to the light. "He must take care. I am no trifler."

It was after eleven o'clock when Mr. Guy returned. Mrs. Harte knew the time to a second.

On the next morning, one of the children happened to be sick, and the doctor was called. In the evening, Mrs. Harte managed it so, that, towards nine o'clock she was alone with Mr. Guy.

"The doctor thinks Frances quite a sick child," she remarked.

"Does he?" Mr. Guy aroused himself from an abstracted state of mind.

"Yes."

"Nothing serious, I hope."

"There is a great deal of scarlet fever about; and she complains of sore throat."

Mr. Guy looked into Mrs. Harte's face steadily, but did not answer. A brief silence followed; then Mrs. Harte said —

"I don't fancy Dr. Blake." The eyes of Mr. Guy had fallen to the floor, but something unusual in the woman's voice caused him to look at her again.

"Has he offended you in anything?"

"No; but he has a prying, inquisitive way about him that I don't like."

"Ah? I haven't noticed it. In what direction does his inquisitiveness run?"

Mrs. Harte did not answer immediately. The question disconcerted her, apparently. But, it was only in appearance. Mrs. Harte was never more really self-possessed in her life.

"In what direction does his inquisitiveness run." Guy repeated the question.

"In a direction by no means agreeable. At his last three visits he has referred to the death of Mrs. Guy in a way which leads me to infer that something is on his mind."

There was an instant change in Mr. Guy's face, and Mrs. Harte noted it well, and took courage.

"What did he say?" The voice betrayed alarm.

"He asked about the quantity of morphia that was given."

"He knew that as well as you or I."

"Perhaps not." Never since the fatal night when he stood, with Mrs. Harte, at the bed-side of his departing wife, had he felt so much in her power as at this moment; never before had Mrs. Harte so meant to make him feel conscious of her power. She came nearer to him, now — nearer, and with an intrusive familiarity that he dared not repel; a familiarity that made him shudder as it approached.

"Perhaps not." Ah, in the tone and manner of Mrs. Harte were something more than in these simple words. It was as if she had suddenly thrown her arms around him, and said, "You are in my power! We are sharers of a fatal secret, and safety lies only in concessions to my will."

"He saw the vial," said Mr. Guy, in a voice which had suddenly grown husky.

"But, it did not give the true indication. I know that three times the quantity indicated by the vial was administered." There was marked emphasis on the pronoun *I*.

"The fact is," she added, after a pause, "I've never felt comfortable in my mind about this thing. My error was in having any part or lot with you in the matter at the beginning. I should have washed my hands clear of it the moment I understood the truth. But," she hesitated, and remained silent for a brief space.

When she resumed, her voice was softer, and she leaned a little towards Mr. Guy.

"But," she continued, "I saw the fearful peril in which you were involved, and believing that no wrong was meant, obeyed my natural impulses, and went over without reflection to your side. The act was imprudent, and I have always so regarded it."

"As Heaven is my witness, no wrong *was* meant," said Mr. Guy, showing considerable disturbance.

"I am sure of that." How skilfully did Mrs. Harte throw just a shadow of sympathy in her voice. "I am sure of that, Mr. Guy." She repeated the sentence, with just a little warmth of expression. "But courts of justice take account only of facts."

"Courts of justice! Madam! What are you driving at?" Guy aroused himself, and drew away from the woman, not able to keep the signs of fear from his countenance.

"Nothing, sir." How calmly spoken were the words. How soft their utterance. There was no stern purpose on her lips, no threat in her eyes. "I merely suggested a fact that no one in your peril should fail to keep in remembrance. An unhappy circumstance — accident, we will say — has placed you in a most unfortunate position, and safety demands that you be always guarded."

"Guarded? — guarded?" Guy's manner showed some bewilderment of thought. "No one is in the secret but you, Mrs. Harte."

You were not guarded when that dropped from your lips, Mr. Guy.

"Those who know me best, sir, will tell you that *I am a warm friend.*"

There was no occasion to add — "*But a bitter enemy!*" for Mr. Guy understood Mrs. Harte to mean that, as clearly as if she had finished the sentence orally — and she meant him to understand it. A shudder crept along his nerves as the conviction grew into assurance, that he was wholly in her power, and that, in the velvet hand which was now laid upon his in a soft, intrusive touch, sharp talons were hidden.

"I can be as silent as death, sir," she said, in her low, unimpassioned voice. "Let Doctor Blake thrust in his probe. He shall find no tender spot answering to his touch. Your secret lies safe with me."

Safe! Mr. Guy felt that it would be safer, if she were stark and cold as his wife lay, when he saw her coffined; and in his heart, he wished she were dead, and out of his path.

"It was only an accident, at worst," he said, endeavoring to rally himself, "and nothing more could be made out of it."

Mrs. Harte looked grave, and shook her head.

"What more could be made of it?" demanded Guy.

"Suppose I were put upon the witness stand, and required to give testimony under oath; what then? The druggist's evidence would be conclusive as to the quantity of morphia sold, and mine would show by what remained in the vial, the quantity given. Any chemist would tell the court that death, in ordinary cases, would follow such an administration. Then look at your position! I tell you, sir, the matter is one to create alarm. I don't like the way in which Doctor Blake asks questions, and shall have to be guarded to the utmost in my answers."

Blank fear was visible in Guy's countenance. Mrs. Harte had narrowed the question of his danger down to a very clearly apprehended point, and he saw his peril more distinctly than it had ever been seen before.

"And you would testify as to the quantity given?" said Guy, looking sharply into the woman's face.

"I would be under oath," was her quiet response.

"And yet, you know as well as I do, that no harm was meant. That, in my anxiety to relieve a maddening pain, I repeated the doses too frequently. And, Heaven is my witness, that I was ignorant as to the effects such small administrations would produce! I never dreamed of anything beyond a long sleep."

"Don't understand me, sir, as questioning this for a single instant," said Mrs. Harte, again laying her soft, cat-like hand upon his arm, with even more of familiar confidence than she had yet assumed. "*I* fully comprehend the case, and you have nothing to fear, unless I should be dragged into court. That is the ultimate result, from which I shrink in fear. An oath, sir, is the most solemn of all obligations."

"And one that you would not violate under the extremest of circumstances?"

Mrs. Harte felt that more was meant than appeared in the words of this question, and, therefore, she did not answer promptly.

"I cannot say what I *might* do in the *extremest* of circumstances," she answered, after a pause. "Human nature is weak. For those who are dearest to us — those in whom life, and all that makes life desirable, is bound up, we often dare a great deal — suffer a great deal — risk a great deal. But, an oath is a solemn thing, and its violation brings consequences that reach beyond this life."

She dropped her eyes meekly, on closing this sentence. Guy studied her face intently. But he was not a skilled physiognomist, and failed to read its signs. After a silence of some minutes on both sides, Mrs. Harte arose and withdrew from the room, satisfied that nothing further was needed to impress Mr. Guy with a sense of the peril in which he stood, and the extent of her power over him. If he had shown indifference to his position — if he had scoffed at her intimations of danger — if he had thrust her back, as she advanced upon him, she would have been in doubt of ultimately gaining her ends; but, he betrayed so fully his weakness and fears, that she felt strong and confident. The image of Mrs. Leslie, as it arose in her thought, when she sat down alone in her chamber, did not now greatly disturb Mrs. Harte. Should that lady really come in her way, she felt that in her own hands was the power of setting her aside — a power in the exercise of which there would be no scruple.

CHAPTER XXV.

EVENTS foreshadowed in the last chapter, took their places as things accomplished in due course of time. At the moment of Mrs. Guy's death, under circumstances that gave her power over Mr. Guy, the thought of using that power to her own advantage entered the mind of Mrs. Harte; and from that time, until at the end of two years, she stood, with orange blossoms in her hair, and heard herself pronounced the wife of Adam Guy, she had never, for an instant, swerved from the first incipient purpose.

Had she formed an attachment for the man, during these two years?— Women often love strangely, and draw, with an instinct of tenderness, towards natures that seem to possess no qualities essential to love. Nothing of the kind! Mrs. Harte's bosom never swelled nor warmed with even the beginnings of affection. If it had not been for the wealth of Adam Guy, and the ends she desired in the possession of wealth, she would have turned from him in disgust, instead of seeking an alliance. As it was, she used him as a stepping-stone to position, regarding him with scarcely more interest than we regard the steps by which we ascend the higher places we seek to gain.

Soon, one by one, disguises fell away from this ambitious woman, and her husband began to comprehend, with a vague feeling of distrust and anxiety, that he had taken an enemy into his household that might prove too strong for him in any war he should attempt to wage. Changes in their style of living were gradually made, almost without consultation; and then, costly articles of furniture purchased with a boldly assumed right of expenditure, that half appalled the man, who still, though possessing large wealth, shrunk from the extravagance indulged by families of far lighter substance than himself. He did not grow liberal as he grew rich; but guarded his coffers with the Argus-eyed fidelity that distinguished him in the beginning. If he ventured on a feeble remonstrance, or even grew earnest and excited over some bolder essay, Mrs. Guy met him with a few calmly spoken and conclusive sentences, that foreclosed argument. Not that he was convinced, but impressed with the futility of offering a word in opposition.

The woman was too strong for him; too strong, because he did not really know wherein her strength lay, nor how to assault it. He still feared her for the secret she held. But for that secret, she had never led him an unwilling man to the marriage altar. It was only because he deemed his good name, perhaps his liberty or life, safest in bonds with her, that he ever permitted himself to be bound; and now, he felt the chains to be very heavy.

Mrs. Harte came into this family, nearly all the elements of which were in conflict from strong selfish proclivities, upon which no wholesome restraints had been laid with no other end but to serve an inordinate social ambition. During the period in which her relation was simply that of housekeeper, she had failed to draw any affection upon herself, even in the younger children; and towards them she had no right feelings. Naturally systematic and orderly, and being moved, besides, by her purpose to win her way by fear or favor to a permanent place in the household, she administered all things appertaining to their external lives in a way that left her without reproach. But, after her marriage — after she took the name and position of wife and step-mother, — a new state of mind naturally led to new action touching the children. They were in her way — six children formed a solid phalanx of obstruction, that would grow stronger with every succeeding year - and the momentous question of how they were to be removed out of her way, came up for consideration, and was deeply pondered. They must not stand between her and final ends — between her, and the absolute possession of Mr. Guy's large wealth, when the time came for him to follow in the path poor Lydia Guy had taken. Did she meditate violence! O no. Nothing of that. But Adam Guy was fifteen years her senior, and she held the chances of survival as altogether on her side. When she became, for a second time in her life, a widow, she desired to have ample wealth as a consoler in widowhood. Having married Adam Guy only for his money, she must not lose the game on which she had risked her all in life.

Philip, the youngest child, had always been ailing. He was puny, fretful and troublesome. Much to the secret gratification of Mrs. Guy — and happily for him — death came in mercy and removed him ere for six months he had tried to say " mother," and feel that the very name was not a mockery to his yearning heart. He died, and the low, soft voice of his step-mother, as she laid her recently jewelled hand on his white brow, said with a tearful tenderness and resignation that deceived only the listeners who were not of the household — " Of such is the Kingdom of Heaven." Truth comes often in hypocritical utterances. It was so in the present instance. Not to find heart-relief in this beautiful sentiment were the words spoken ; but to give an impression of religious faith and maternal affection where none existed.

The birth of a son to Mrs. Guy was an event on which momentous issues depended. If, before this period, she had pondered the virtual disinheritance of her predecessor's children, the accomplishment of that result became, now, a well-defined purpose. The motive was strengthened seven fold. Already she had begun her work in a secret fostering of antagonisms that existed among the children, so that permanent alienations might result, and the enemy she had to encounter stand divided and in conflict with itself. After the birth of her own child, Mrs. Guy used every means within her reach to turn the father's mind away from his other children. She gave him no peace until Adam, John and Lydia were sent away from the city to a boarding school, where they were kept for years, only visiting their homes during seasons of vacation.

With the pleasure that only an evil spirit could realize, Mrs. Guy saw at each return of the older children, that John and Lydia were moving in paths, the end of which would be almost certain alienation from their father. John's spendthrift habits, and animal propensities she stimulated by frequent additions to the limited supply of pocket money that was allowed; and he had learned to write to her in confidence, and solicit these additions with a certainty of always receiving a favorable response. In sending him money, Mrs. Guy never failed to give admonition of the soundest kind, and always enjoined secrecy, as the knowledge by his father of these departures from his wishes, would result in cutting off all indulgence. John loved himself and the means of enjoyment thus conferred, too well to betray their secret; and as habits of self-indulgence grew stronger, his calls for money increased, until Mrs. Guy found the drain upon her individual purse, becoming far heavier than she could conveniently bear. The limitations had to be imposed. Against limitations sensuality always rebels. The boy demanded increased supplies; but his step-mother was, whenever it suited her so to be, as unyielding as iron. She did not meet his demands for an increase, but lessened the sums she had been accustomed to transmit. Self-denial was not in the boy. Towards restriction, he had only one course of action, and that was resistance. But, resistance to his step-mother he soon discovered to be a waste of strength. He must take from her just what she chose to accord, and make up the balance of his wants in some other way.

In some other way! Ah, there was peril here! Deeply versed in human nature, Mrs. Guy was not unaware of the boy's danger. She knew very well, that the good advice, about curbing appetite and desire — about self-restraint, and all that — which she urged in her letters, would pass him like the idle wind, and that he would cast about for some means of obtaining the sums of money which she denied. Her indulgence had given his grosser propensities too large stimulus, and to the cutting down of gratification he would not submit. If one source of supply were diminished, he must find another.

Mrs. Guy had not been well informed as to John's defect of principle, but many lapses from integrity had come under her own observation, and she was, therefore, fully prepared to hear, at any time, of his disgrace at school, on the accusation and proof of dishonest appropriations; and not only prepared as to an anticipation of the fact, but strong in a spirit of submission and resignation whenever the fact should be announced.

Several years had glided away since the three older children left home for school. Adam had gained his nineteenth year; John was reaching towards eighteen and Lydia was a tall, womanly looking girl, only a year younger than John. At home, the family had been increased by an addition of two more children, numbering three born to the second Mrs. Guy — two sons and one daughter — all living arguments against the rights of Mr. Guy's first children.

At this period, we will bring the reader, for a little while, to a nearer point of observation, and let him see in what measure the ends of life with Adam Guy, are working out the grand results of happiness which, twenty years before, made the future look so bright with promise, that patience to wait the slow evolution of years could hardly be maintained.

Mr. Guy left his home one morning, with the pressure on his feelings that always rested there while at home — a pressure which had its origin in an ever abiding sense of weakness. Out in the world his money was an undisputed argument. His will was free; his word a law. But, at home, as in the latter days of his first wife, there was a subtle, almost intangible power, against which resistance seemed hopeless. If he struck against it, like a man beating the wind, the result was only self-exhaustion. Mrs. Guy did not, like her predecessor, oppose an open resistance, or stifle him with impassiveness. She was usually calm, self-possessed, and gentle in tone; never meeting him with resolute opposition; yet, was he all the while conscious, that she was bending him to her will, and gaining at every point of approach. She was the watchful spider, silently spinning her web. He did not feel the silken cord, of invisible fineness, as it fell lightly over him, and only knew that he was in her toils, when some movement warned him of his bands and his helplessness.

Mr. Guy left home one morning, as we have said, with the usual uncomfortable weight upon his feelings, and repaired to his store. A number of letters were on his desk, most of them business letters, directed to the firm. Among them, were two addressed to himself. One of these, he recognized as from his son Adam. The business letters were first read. Then he broke the seal of one of those that remained. It read thus: —

"DEAR SIR: — I have most painful intelligence to communicate. Your son John has been guilty of conduct that renders it impossible for me any longer to continue him in my school. The charge of appropriating money, and various articles belonging to other persons, has been so fully proved against him, that no question as to his guilt is entertained by any one. Not wishing, for your sake and his own, to subject him to the disgrace of expulsion, I write to request that you will direct his immediate return home. He is already suspended from participation in the exercises of my school.

"It is also my duty to say, that John's habits are of a dangerous kind, and will, unless they can be broken, lead him to certain ruin. The freedom with which he has been supplied with pocket money, has led to constant self-indulgence, and he has, during the last six months, as I now learn, been several times intoxicated. To my surprise and pain, I have been informed within twenty-four hours, that he has, for some time, kept wines, and other liquors, in his room. I very much regret that a knowledge of this fact was so long concealed from me by those who were advised of it. My object in communicating it now, is that you may clearly comprehend his danger, and so be able to adopt the most effectual means for his rescue.

"With high respect, I am, truly yours,

The first reading of this letter, so stunned Mr. Guy, that his mind fell into a state of painful confusion. He rallied in a little while, and read the letter again, when the whole truth stood out in all its shocking magnitude.

"A thief and a drunkard!" Mr. Guy shuddered inwardly, as he repeated this sentence to himself. He was in his counting-room, with clerks around him, and must not betray an outward sign of the agitation against which he was struggling.

Next, Adam's letter was opened. It recounted, with some particularity, the criminal conduct of his brother, condemning him in strong language, and prophesying no good in the future. "He doesn't seem to have a single redeeming quality," was the strong language used by this boy, "after his father's own heart," followed by sentences like these:—"He only thinks of self-indulgence, and would spend a thousand dollars a month, if he had his will." "I don't know how it is, but, to my certain knowledge, he spends three or four dollars to one of your allowance. Where does the rest come from? He's in debt to the boys; but not enough to account for this. I'm afraid he gambles." The letter closed as follows:—"I must come home, also; for, after John's conduct, I can't look an honest boy fairly in the face. Any way, I'm tired of study, and want to get into business."

When Mr. Guy went home at dinner-time, he said to his wife, in a tone that betrayed his unhappy feelings:—

"The boys are going to leave school."

"What?" Mrs. Guy turned upon her husband with a suddenness of manner, and a degree of surprise, unusual for one so guarded, and so externally placid.

"The boys are coming home from school."

"To remain?"

"Yes."

"Why?"

"John has got himself into some trouble, and Adam doesn't wish to remain after his brother leaves."

"But haven't you a word to say on a matter like this?" demanded Mrs. Guy.

"The matter is pretty well out of my hands," returned Mr. Guy, with some impatience.

"Out of your hands? I don't understand you."

Mr. Guy had not intended to show the letters he had received to his wife, for he was not ignorant of the fact that she was more ready to defend John than to blame him; but, acting under a confused impulse, he drew the teacher's communication from his pocket, and placed it in her possession. She read it through, calmly.

"This is trouble!" fell from her lips; but, her voice did not add to the force of her words; for, already, she was considering the facts revealed with reference to their bearing on her future schemes.

"Trouble, and disgrace added!" said Mr. Guy, with a stormy vehemence of manner, that sometimes betrayed itself under unusual provocation —"I'm tempted to disown the wretch! A son of mine turn thief! Ugh! Horrible!"

"I would sooner see him dead," answered Mrs. Guy. How closely her declaration came to the actual truth.

"Dead! Ho! Death would be as nothing in comparison."

"What are you going to do with him?" Mrs. Guy put the question almost sharply, her interest in the matter betraying her from the citadel of her strength — external calm.

"Do? Heaven only knows!"

"Idleness in a city will only make ruin the swifter. Temptations meet the unwary at every step."

"I know — I know. He must be set to work at something," answered Mr. Guy, casting about in his thoughts, but without seeing any light.

"I wouldn't bring him home," said Mrs. Guy, speaking from the over anxiety she felt in reference to the two boys' return.

"Why not? What would you do?" Guy knit his brows, and looked sternly at his wife. She had betrayed herself, and he saw a little below the surface. "Isn't home the best and safest place for him? — Home, where father and mother can watch and guard, warn, lead, or admonish? I know that temptation lurks in cities; but home influence ought to be stronger than temptation. Let us see if it cannot be made so in John's case."

"*I* can promise nothing," answered Mrs. Guy, drawing coldly back into herself. "As for Adam and John, they have always acted with as much independence as if I were a nonentity. They have never clearly acknowledged my rights in the household; and, were I to attempt control or influence, so far as they are concerned, open war would be the consequence. There is no use in concealing the fact, John is wholly beyond my reach;

and if, in bringing him home, you calculate anything on my power of restraint or direction, you build on a foundation of sand."

Mr. Guy did not answer. What could he say? The will of a woman like his wife, was too strong a thing for him to act against; particularly, when to the will was added a subtle and far-seeing spirit. He did not venture to speak of duty, forbearance and self-denial; for these had never been elements of power in his own life; and he doubted as to their existence, as moral forces, with any one. They might answer as catch words, and to make oratorical points in a sermon; but were of little worth as ends of action — certainly of no value in his home. No; he did not speak to his wife of duty, forbearance, or self-denial, lest she should fling the words back in his face with cold contempt.

"If Adam and John come home," resumed Mrs. Guy, Edwin must go away to school somewhere. He's nearly past me now, and, in league with his two older brothers, will set my authority aside, as nothing. The fact is, Mr. Guy, the house isn't large enough to hold us all; and you might as well comprehend the fact first as last. Three to one are too many; and I can't make my way against such odds. With Edwin alone, I am taxed to the utmost to maintain myself in peace. Put Adam and John on the enemies' side, and I shall be driven from the house in less than two months."

"You speak wildly," said Mr. Guy, in a tone of impatience, beginning to stalk about the floor in quick, short turns.

"Not wildly but soberly. Facts are stubborn things,

sir. You don't know half of what I'm required to put up with from your children, old and young. They are the curse of my life!"

Mr. Guy stopped suddenly, before the woman who said this, and gazed at her with a countenance on which surprise blended with a shade of fear.

"That is strange talk," he ventured to say.

"And stranger that it is true talk," answered Mrs. Guy, firmly. "No woman could have done more than I have done for your children's well being and comfort; but, they set themselves against me from the beginning, disputing every inch of ground I assumed in the family, and yearly gaining strength, until now, they are able to set me at defiance. If you go over to them, the end has come. I must step aside, and find protection somewhere else."

"Woman! Are you beside yourself?" exclaimed Mr. Guy.

"No, sir," was the cold, steady reply. "I am in possession of all my faculties, and able to comprehend, clearly, my position. The odds are against me. When Adam and John return, I shall, in all probability, have to take my children and seek another home. Submit to dictation and insult from them, I will not! Enough has already been endured."

There was a flashing light in her eyes, never seen by Mr. Guy until now. Hitherto calmness of purpose had marked all her actions. If she set herself in opposition to her husband's will, she wrought so in concealment that he had scarcely a suspicion of her purpose until it was accomplished. But now, she stood revealed in a sterner, more resolute, and more defiant character. She meant to come into open conflict with his elder children — not recklessly and blindly ; nor in any doubt as to the issue, — but well assured in her own mind of victory.

"But, what am I to do with them ?" asked Mr. Guy. "They are my children, and this is their home."

"Require their just submission," said his wife firmly.

"Easily said," was replied, with returning impatience.

"If you cannot rule them I cannot," answered Mrs. Guy, "and so the antagonism remains. As for me, I am a lover of peace and order; but, at the same time, will not accept peace at the cost of humiliation. Self-respect forbids that. As your wife, and their mother by your election, I will never submit to their insults, defiances, and impertinences. While they were little children, I bore everything, as was my duty, trusting to win their love ; but now, when they are on the verge of manhood, I am absolved of duty, and will stand upon my rights — asking nothing and yielding nothing. If Adam and John are to come home, well. But unless you send Edwin away, there will be no peace or safety. Adam and John are always in opposition to each other, and I may stand between them, giving a side to each, and so maintain myself; but if Edwin stays at home, he will go over to one or the other, and hold me at defiance. Therefore, think well before you act, for I am in earnest. I trust there will be no open rupture with me and the boys; but if they attempt to set me aside now, it will as surely come as the night comes after the day. So, I pray you, act with circumspection. There is a crisis at hand."

CHAPTER XXVI.

MR. GUY was a very shrewd merchant, and none was wiser than he in all that appertained to the making and keeping of money. But, take him outside of money-schemes, and he was shorn of his strength. As a money-maker, he was great; as a man, nothing. Present to him a question of trade, or finance, and nine times in ten his decision would be of the soundest character; but, let the question involve political expediency, or social law, and he had no skill — no perception. He comprehended the operations of business thoroughly, and understood human nature on the business side; but mental and moral movements puzzled him, and human nature on the social side, was a mystery he had no skill to penetrate. With him, it was, literally, nothing but money. All his wisdom lay crystallized around his love of gold.

Mr. Guy was not, therefore, equal to the new position of affairs in his family. He was adrift on a troubled sea, without chart or compass. The resolute attitude of Mrs. Guy, confounded as well as confused him. He knew her well enough to be assured that, in assuming this attitude, she had changed her front for the purpose of strengthening, in some way, her position; and that she would strengthen it, and so gain some new advantage, he felt sure — the conviction oppressing him with a sense of his own weakness in her hands. She was a power, acting upon him in such subtle and strange ways, that he could make no sure defence. Usually, if he threw up a barrier, and entrenched himself, the enemy retired, leaving the post of no value, but weakening and annoying him with assaults from unexpected quarters. Now, however, she was bearing bravely down upon him, with all her banners displayed, meaning to risk a battle. Was he strong enough to meet the shock? Did he know the strength and resources of his enemy?

Mr. Guy's heart failed him. The attitude of his wife was too bold; too full of conscious strength; too resolute. The time had come, when he must choose between her and his older children. If they did not submit themselves to her, they, or she, must go out from his home. That was the alternative clearly offered. A feeble attempt at remonstrance and persuasion was made, but Mrs. Guy turned it aside as futile. She would have no parley.

"I ask nothing, sir, but what is right and just — nothing that you should not, of your own motive, secure to me as your wife,"— was the tenor of all her answers. "If I am to be insulted and set at nought here, to whom, pray, are you to look for the integrity of your home? Let your sons deport themselves in a becoming manner, and all will be well. If they do not, the responsibility of what follows is with you and them. I shall wash *my* hands clear of all stain."

Adam and John were not at home a week, before the storm hung dark above.

Their father had warned and admonished them faithfully; but they were not of those who profit by warning and admonition. Passion, prejudice, appetite and self-will, were, one or all, their counsellors. In order to get Adam out of the way of his mother, and the temptation to annoy her, Mr. Guy gave him a place at one of the desks in his counting room, and kept him fully occupied all day. This was a highly conservative movement; and if Edwin, now fifteen years of age had been sent away to school, Mrs. Guy, by a system of demoralizing indulgences, could have maintained the balance of peace with John, until in his steadily progressing downward course, he reached a point of depravity at which his father would cast him off. In this way, sooner or later, Mrs. Guy saw that John would be disposed of, and thus out of her path. But, Edwin's presence at home, united as she had prophesied, the forces against her, and she began to set her own in battle array. In this she was the superior strategist, and wrought silently, and in secret, until the time for opening the contest had arrived. Non-combatant she appeared to be, in the eyes of John and Edwin, who made one advance upon another against her authority, she seeming to yield as they advanced, until she had them completely in her power. Let us see how the position stood.

John's humiliation at school, did not make him shamefaced at home. He had lost, even at his early age, all sense of honor. Appetite and passion only grew more clamorous from restriction, for his animal nature was in the ascendant. His father cut off all supplies of pocket money, this being the only way to punish and restrain that he could devise; and, very naturally John applied to his mother. Instead of meeting his applications, as in former times, Mrs. Guy said — "Go to your father," — thus pushing him away from, instead of drawing him near to her, and conciliating him by indulgence. So John put himself in opposition, and, out of a revengeful spirit, assailed her, on all fair occasions, with annoyances and disrespect. These she bore with a quietness that encouraged John, to whom Edwin, a weak, but not naturally vicious boy, went over, and grew bold in contemning his mother's authority.

"John," said Mrs. Guy, one morning, about ten days after his return home, "go to my room and bring me my purse. You will find it in the left hand, small drawer, of my bureau."

"Give me the key." John was advancing towards her ere she was half done speaking. Only three minutes before, he had refused to get her a book from the library.

"The drawer is not locked," was answered.

With a springing step John left the room, and in a few minutes returned with the purse.

"Thank you," said Mrs. Guy, as she took it from his hand, and placed it in her pocket. She had no need to examine its contents to be assured that John had helped himself. Eyes like hers read faces as if they were books.

In less than twenty minutes John and Edwin went out together and were gone all the forenoon. Mrs. Guy knew to a penny how much the purse contained when it came into John's hand; for she had counted the half dollars and bills over twice. Two half dollars and a five dollar bill were missing. A gleam of satisfaction went over her face as this fact was ascertained, and she said to herself, speaking aloud —

"As I expected."

It was nearly dinner time when John and Edwin came home. John was self-possessed, and rather jaunty; but Edwin's face wore a shy look, and there was an air of embarrassment about him that did not escape the keen eyes of his step-mother. Purposely, she drew near the two boys, so as to get their breaths in speaking; and discovered, what she had already suspected, that they had been drinking some kind of intoxicating liquor. The fact did not shadow her placid brow.

John was quieter than usual at the dinner table. Occasionally, Mrs. Guy detected a look of inquiry, sent across to her half covertly. John was, evidently, in uneasy debate on the question as to her knowledge of his guilty inroad on her purse.

The first course had passed, and they were nearly through with the dessert, when Mrs. Guy, without preliminary, or warning, asked, looking at John —

"How much money did you take from my purse this morning?"

The suddenness with which she put the question disconcerted John. His face grew red, and there was some hesitation of manner, ere he responded, in an angry, repellant tone, —

"I don't know what you mean."

"You went to my drawer this morning?" said Mrs. Guy, without the slightest sign of weakness in her even tones.

"You sent me to your drawer," asseverated John, vehemently.

"I know."

"You asked me to bring your purse. Edwin was present." And John glanced towards his brother, whose pale face betrayed his knowledge of, and participation in wrong.

"Very true; and you brought it. My question referred to the sum you took therefrom before delivering it into my hands."

"Not one dollar!" said John, — angry and positive; and he offered a startling oath in confirmation of his denial, shocking and astonishing every one at the table.

All this while Mr. Guy had remained silent, like one half stupefied. Now he aroused himself, and in a loud voice, looking at John, cried out, —

"Silence, sir! How dare you use such language here?"

"I dare anything when falsely accused, sir," answered the boy, boldly.

"When my purse was taken into your hands, it contained twenty-four dollars; and when you placed it in mine, the contents were reduced to eighteen," said Mrs. Guy, speaking slowly but firmly. "Ten minutes before you went for my purse, you refused to get me a book from the library; but, when the request to bring my purse was made, you were off with a spring. It is useless for you to deny the fact of taking six dollars. I saw it in your face, as you handed me my purse. In a little while afterwards, you went out with Edwin, and were gone all the morning — spending the money of course, and not, as I have satisfied myself, in the best and safest way. The stale fumes of an oyster cellar were in your clothes, and the smell of liquor on your breaths. I noted the fact well."

John alone, might have braved his stepmother out, in positive denial; but, the tell-tale face and manner of Edwin, turned his father's attention to him; and a few sternly put queries brought out a clear confession of the truth.

"There is only one safe thing to be done," said Mrs. Guy, to her husband, when they were alone. "Edwin must be saved from the ruin into which John will certainly drag him, if they are left together. Edwin is weak and easily influenced. Since John came home, I see a change for the worse going on daily."

"I shall send him away to school," was the positive answer of Mr. Guy. "He must not remain at home for a single week longer."

And he did not. Just as Mrs. Guy had planned, the event came out. Edwin was sent away from home, and kept as liberally supplied with pocket money by his step-mother, as John had been. Unhappily for the future of the boy's life, the effect was just what Mrs. Guy designed that it should be. Temptations spread themselves all along his unguarded path, and his feet were ever wandering.

CHAPTER XXVII.

NOTWITHSTANDING the threatening attitude assumed by Mrs. Guy — notwithstanding the marshalling of her forces — it was no part of her plan to risk a serious battle, if that desperate issue could be avoided. By a resolute bearing she made strategy the more successful. The fact, that Mr. Guy believed her when she said that she would leave his house rather than permit his children to exercise a dominant influence, caused him the more readily to fall in with her wily plans for removing them to a distance.

For awhile John held himself coldly aloof from his stepmother; but she, after Edwin's departure for school, gradually broke down the wall of angry reserve which he had thrown up between them, and assumed a degree of interest in the boy that, with a little indulgence in spending money, laid his mind open to almost any influence she might choose to exercise.

"This idle life, for one of your age, is dreadful, Jonn," she said to him, as they were alone, one morning, speaking in a tone of interest.

"Oh, well, father's rich!" he answered, tossing his

head in an independent, don't care sort of way — "there's no use in my doing anything. A gentleman's life for me."

"For shame, John! Anything but a drone or an idler. Adam goes to the counting-room every day."

"Adam! Pish!" And John curled his lip. "He'd eat dirt for a dollar, and then bury the money after it was earned."

"Adam loves money too well, I know," said Mrs. Guy; "but he's willing to earn it."

"Don't quote him to me," returned John, with some impatience, "the mean, stingy fellow! Let him earn his money and keep it, if he will; but don't hold him up as my exemplar."

"I'll tell you what I've been thinking, John," said his stepmother, changing her tone, and speaking in a way meant to inspire him with the thought in her mind. "You'd like to see the world, I know. Every young man does. Your father's firm is loading a vessel for the Pacific. Now, why not go in her as supercargo?"

John started up, and stood, all interest for a moment or two; then clapping his hands together, he replied, while a warm flush came into his face —

"I'd like that! But —" And his countenance changed a little.

"What?"

"Father would say no."

"I'm not so sure of that, John. But, would you really like the place?"

"Of supercargo?"

"Yes."

"Grandly! I've always had a wish to go to sea." There was an eagerness in John's manner that showed how strongly the idea was taking hold of him.

"I do not believe that your father, on reflection, will object," said Mrs. Guy.

"Will you ask him about it?"

"Yes. And I think you'd better leave the matter entirely in my hands. And, first, let me caution you not to say anything on the subject to Adam. He has considerable influence with your father, and would be sure to oppose, if only for the sake of opposition."

"Oh, I understand that! He'd thwart me out of sheer malignancy."

Mrs. Guy did not throw in a mollifying word. It was no part of her mission in the family to harmonize or conciliate.

"I will manage him," she said, in the tone of one who felt her power; "only, as I suggested, keep your own counsel. You shall go as supercargo in the Ariel if you desire it."

"I do desire it above all things," replied John, "and if you will get father's consent, I'll remember you as long as I live."

"There is one thing I would suggest, as your father will have to be managed a little in the beginning," said Mrs. Guy, lowering her voice in a confidential way, and speaking with an unusual familiarity, "and that is, an assumption of indifference on your part. This indifference, may have to take the form of opposition before all is settled; but, I will give you the right hint at the right time. You understand?"

11.

"Perfectly."

"It wont strike him favorably on the first blush; but I'll engage to bring him over to our way of thinking. It will be a splendid chance to see the world and improve yourself, and when he understands this there will be no more trouble."

Mrs. Guy lost no time, for the Ariel was already taking in cargo, and would be ready for sea in less than two weeks.

"This idle life that John is leading, troubles me continually," said Mrs. Guy, thus opening the subject, on the next occasion of being alone with her husband. "It will be his destruction, I fear."

A shade of anxiety passed over Mr. Guy's face, but he did not respond.

"Is there nothing that we can do with him?"

Mr. Guy shook his head.

"He'll be ruined if left to his own will — hopelessly ruined."

The voice of Mrs. Guy was full of concern. Still, her husband made no answer.

"How would it do to send him out as supercargo in one of your vessels?"

"It wouldn't do at all," was Mr. Guy's quick answer.

"I'm not so sure of that," said his wife, who had expected just this answer. "If he could be induced to go, it might be his salvation."

"Do you suppose we'd trust a boy like him with the disposal of a cargo? What does he know of business?"

"Of course," answered Mrs. Guy, not in the least disconcerted, "you would send him with an experienced captain, who would be the real man. The end is to save the boy; to get him away from the temptations that now beset him on every hand. As to his control of the cargo, that is another thing. He might sail under the idea of full powers, while the captain had orders to supersede him on the ship's arrival out. Don't you see, how safely this might be done? My only fear is, that John may not consent."

"He will not; on that you may rest assured," said Mr. Guy.

"If he were to consent — what then? How does the thing strike you, on reflection?"

"Anything to get him away from the dangers of this city."

"So I think. Turn it over in your thought, Mr. Guy. How in regard to the captain of the Ariel? — Is he a discreet man?"

"He is a good captain," was replied.

"Does he need a supercargo?"

"No. The ship will be consigned to a house in Valparaiso."

"So much the better. John's position as supercargo would give no control whatever, and might be so arranged with the captain as not to embarrass him in any respect. Take the captain fully into your counsels, and let him manage John in his own way after he gets him to sea."

"You talk as if the whole matter was settled," said Mr. Guy, not able to repress a tone of impatience.

"And why not, if right to be done?" was coolly answered. "You can make all fair with the captain, of course. That part is easily arranged. The serious difficulty in the way is to get John's consent."

"A thing not to be hoped for, in my belief. John is too fond of ease and self-indulgence, to risk the possible hardships and privations of a trip around the Horn."

"It will require some management." This, Mrs. Guy admitted. "But, if you can arrange with the captain to let John go as nominal supercargo in the Ariel, I will undertake to bring him over to our wishes."

"I may safely promise my part of the work, then; for I have no faith in the accomplishment of what you propose."

"But, Mr. Guy," said his wife, with increasing earnestness of manner, "don't you think it would be of great use to John?"

"Perhaps it might be; but, there's no telling."

"It would remove him from temptation."

"Yes."

"And bring him under rigid discipline."

"I don't know about that. A supercargo is not a sailor, nor even subordinate to command like a ship's officer."

"Oh, as to that, you and the captain could understand each other. The object in sending him to sea must not be forgotten. I think, maybe, it will be just as well for you to hint the matter to John yourself, and see how his pulse beats. Perhaps he may be carried away with the idea as a novelty, and so all run smoothly. But, don't urge the matter, if he object, or seems to consider it of any importance. He will be sure to say something to me about it."

"And go dead against the whole thing, should you favor it in the least. I know him."

"He's not apt to yield in favor of my plans, as I'm aware," said Mrs. Guy. "But, where there's a will, there's a way, and I'll undertake to manage him."

The more Mr. Guy thought over his wife's suggestion, the more in favor of sending John to sea did he become. As he dwelt on the subject, a hope for the boy kindled in his mind — a hope that love of business and gain might be stimulated. A small adventure of his own might be entrusted to him, with direction to invest the proceeds in merchandise for the return voyage. This view was dwelt on, until it looked so promising that Mr. Guy believed in its power to save his son. If by any means, a love of gain could be stimulated, he felt sure that all would be well. He had an undying faith in money. It was, in his thought the only conservative power. No concern about Adam's future haunted his mind; for the boy's love of money was, in his regard, a sure protector.

John played off and on with his father, according to the programme of his stepmother, and affected to yield, finally, with a great deal of reluctance, and only after securing sundry privileges and advantages, which, in the beginning, Mr. Guy never thought of conceding.

There was an unusual lightness in the heart-beat of Mrs. Guy, on the day John departed in the Ariel. She had an evil faith in the result of his voyage. If he ever came home at all, which might not be, she believed that he would come home so much worse in morals and habits, that no hope for his manhood would remain ; and so, he would cease to stand in any formidable manner between her and her ambitions. At any rate, he was out of her way for a year; and she took all beyond that time on trust.

CHAPTER XXVIII.

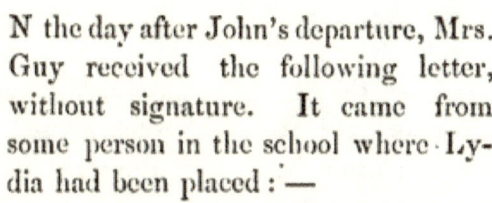

ON the day after John's departure, Mrs. Guy received the following letter, without signature. It came from some person in the school where Lydia had been placed:—

"MADAM:—I feel it my duty to say that your daughter Lydia is receiving the attentions of a young man in this neighborhood, who cannot possibly be acceptable to her family. To my certain knowledge, they hold clandestine interviews at night, she stealing from her room at a late hour, to join him. My concern for her welfare, prompts me to send you this information."

Twice did Mrs. Guy read this communication, without exhibiting a sign of disturbance. Then, taking a match, she lighted the gas, and holding the letter in the flame until it was consumed, flung the charred flakes from the window.

The letter was anonymous, and behind this fact Mrs. Guy shielded herself. That it gave her the exact truth, she did not question for an instant; and yet, speaking in her external thought, she said, as the fire devoured the paper—

"The mean accusation of some jealous girl, afraid to sign her name."

Yet, she was sure in her heart it was not so; sure in her heart that Lydia was in peril, and should instantly be brought home. A week passed, and then came another letter, written by the same person, but without signature.

"Fearing," she said — the writer was a woman — "that a communication sent to you several days ago, may have failed in reaching its destination, I address you again on the subject of your daughter, now at school in this place. Do you know that she is receiving the attentions of a young man residing here, and that she is in the habit of meeting him at night, clandestinely? The young man is well enough in his way, but not the one that you could accept as Lydia's husband. Pray look to the matter before it is too late! I have now, twice, given you warning, and so washed my hands clear of the whole matter."

"Anonymous!" And Mrs. Guy shook her head. A lighted match — the gas in a flame — and then a little handful of black ashes were flung from the window.

"It wont do, my jealous young lady." A cold smile played over the lips of Mrs. Guy, as she sat down, and took up a book that she had been reading.

With an eager interest, that absorbed nearly every thought, did Mrs. Guy wait the issue which she believed to be at hand. If Lydia should contract a clandestine marriage with a young man having neither social position nor wealth, the anger of her father would know no bounds. He would disown and disinherit her without remorse; and so, she would be out of her stepmother's way. To make this separation permanent, would be an easy task to so clear-seeing and unscrupulous a woman as Mrs. Guy. She had not very long to wait. One forenoon, less than two weeks from the receipt of her last anonymous letter, Mrs. Guy heard the street door open, with a rattle of her husband's key in the lock, followed by his heavy tread, quicker than usual. He called her, as he came up stairs, and she answered from her chamber, where she happened to be.

The forewarned heart of Mrs. Guy guessed truly the meaning of this unusual appearance of her husband. His face was pale and agitated, as he entered. An instant, he glanced around the room, and seeing that his wife was alone, shut and locked the door. Then drawing a letter from his pocket, he thrust it into Mrs. Guy's hand, saying, in a desperate kind of way —

"Read that!"

As he stalked about the room, like an animal smarting with pain, his wife in her unruffled way, unfolded the letter, and read—

"To Adam Guy, Esq.

"Dear Sir:— The thing's done, now, and there's no helping of it. I married your daughter last night, and she's my wife forever. I love her as my life, and all for herself alone. I hope you will not be angry sir. I couldn't help loving her. We were afraid you wouldn't consent, and as we couldn't live without one another, we took the risk of getting married, being sure that when you saw how we loved one another, and couldn't live apart, you would forgive us. Dear Lydia wants me to write first. She will write to-morrow.

"Affectionately, your dutiful son,

"JAMES BRADY."

"Well, that's a nice business, upon my word!" ejaculated Mrs. Guy, showing considerable feeling. "Is the man a fool, and the girl mad?"

"Confusion! Curse him!" Mr. Guy threw the words out with a raging force, his eyes in a flame, and white spots of foam on his lips. "Had you any suspicion of this, madam?" He glared upon her like a devouring beast.

"*I?*" She drew her form to its utmost height, with an air of supreme astonishment. "I, sir? What do you mean by such a question?"

"Had you no suspicion of this? I speak plainly, don't I?"

Mrs. Guy stepped back a pace or two from the half madman who confronted her, yet without removing her eyes from his distorted countenance.

"The words are plain enough," she said, in a steady voice, the coldness of which gave a chill to the hot blood of her husband. "But, I am yet in doubt as to their full meaning. Perhaps, in this excitement, you have forgotten who I am."

Mr. Guy was never strong enough for this woman, when she set herself against him. In all the contests which had occurred, she manifested such a resolute spirit, and showed such a consciousness of possessing any amount of reserved force, that he shrunk from a desperate trial of strength. And so it was now, for he did not again repeat his accusing interrogation, but worked down his excitement, by pacing the floor rapidly. Stopping, at length, and confronting Mrs. Guy, who had resumed the seat from which she had risen, he said, with bitter emphasis —

"I disown her from this hour! I cast her off utterly! She shall be to me as one dead! As for the man, I will spurn him as a dog, should he ever cross my path."

"She is still your child," said Mrs. Guy.

"Silence!" The fire flashed out in a sudden gleam. "Don't cross me here, madam, or there'll be trouble between us! She is my child no longer. A beggar's wife, let her live and die a beggar, for all I care."

"It's a hard thing to bear. What could have possessed the girl?" Mrs. Guy dropped in these sentences skillfully.

"The devil possessed her," said Mr. Guy, brutally.

Mrs. Guy covered her face with her hands, and actually expressed a few tears."

"None of that with me, madam! It wont do. The girl has made her bed, and she'll have to lie in it, even to the end.

"She is so young," suggested Mrs. Guy.

"There! there! None of that, I say!" Mr. Guy spoke with angry impatience.

Enough for appearances had been advanced by this designing and cruel woman; and so, she said no more, but let her husband's indignation have free course.

It soon became apparent to Mrs. Guy that this act of Lydia's had touched her father very deeply. If he had felt regard for one of his children more than for another, Lydia might be called the favorite; and she was not to be cast off utterly without pain. She noticed an unusually drooping forward of his head, as if a weight were resting on his shoulders; and a severe abstraction of manner, from which, if he was disturbed, he came out with an unchecked impatience.

Two days after the letter from Lydia's husband was received, one came from Lydia herself. The post-mark, and hand-writing in the direction, indicating its source, Mr. Guy, without breaking the seal, enclosed it in an envelop, and sent it back, unaccompanied by a word.

Two weeks had passed. Mrs. Guy sat in the midst of her own children, three in number, with her thought dwelling in their future, which she resolved to make sunny and prosperous, no matter what other skies were darkened, or what other rights sacrificed, when the door was flung open, and Lydia came hastily towards her, across the room.

"Oh, mother!" fell eagerly from her lips.

But Mrs. Guy lifted her hands in a repulsive attitude, and partly turning away, said icily —

"Don't come near me!"

"Oh, mother!" repeated Lydia, in a choking voice, stopping midway of the room.

"Don't say mother to me. I am not your mother!" Lydia had never seen, in the face of her step-mother, such malignancy and hate as now smote upon her. All disguise was thrown off; as she felt, so she looked — a cruel and unrelenting enemy. "You have dug an impassable gulf between us, and it will lie there forever. Go from this house! You have neither part nor lot in it. It is your home no longer. A wicked, disobedient child, must take the punishment that is her due."

A few moments Lydia stood in a kind of maze. Then a wild look of despair swept into her face, as if suddenly conscious of a new and fearful condition, from which escape was hopeless.

"Go!" The right hand of her stepmother waved, in imperious enforcement of the command.

"I must see my father," said Lydia, rallying herself, and speaking with some firmness of tone.

"Oh, very well!" replied the stepmother, mockingly. "See him, if you will. Call in the evening, when he is at home."

"I will wait until then," said Lydia.

"Excuse me, no! You cannot wait until then. My word is law here; and I say that you cannot remain for even one hour under this roof. So, take your departure!"

Literally staggering back, in a sudden weakness, from the wolfish eye of her stepmother, Lydia went from the apartment. Mrs. Guy followed, to see that she left the house, and literally pursued her to the very street door.

The poor misguided child had come alone, to the city, in order to get reconciled, if possible, to her father and stepmother. What of her marriage? Was there any hope in it? Any basis of character or moral quality in the young man to whom she had given in trust the well-being of her life? Not much basis, we regret to say. He was, however, rather weak than wrong; and under right external influences, would have made a man of ordinary standing — good enough in his way, but of no force in society. His education was defective, and he lacked both the ambition, and mental activity, which lead to self-improvement. The easiest way for a young man to get along in the world, who has not the ability to advance himself, is to marry a rich wife. At least, such is a current opinion. Young Brady chose this method, and finding in Lydia Guy a fair subject for conquest, set himself to the task of winning her favor. He understood some of the arts to be used, and was successful. Lydia fell into the snare laid for her feet, and forgetful of prudence and duty, cast the fatal die that changed, in an instant, her whole relation to society. Alas, poor child! Not for herself had she been wooed and won; but for the money which her sordid and mean-spirited suitor believed would accompany her hand. She had come alone, as we have said, in order to effect a reconciliation. The distance from the town in which she had been going to school was over two hundred miles, and it had taken all the money that remained in her possession, to pay the expense of getting home. Lydia had expected anger, harsh words, and stern rebuke; but was not prepared for an absolute expulsion from her father's house. No wonder that she staggered away from the door with unsteady feet; nor that people turned and looked after her, strangely. Where was she to go? It was over two years since she left home for school, and in that time, girlish friendships in the city had died out.

Moreover, the selfish isolation in which her father had chose to live, had so circumscribed their friends and neighborly relations, that Lydia was little more, at this time, than an unknown one, in the place of her nativity.

Where then, was she to go? Alas, for the poor child! there was not, in all that great city, a single door at which she might enter, in confidence of a welcome such as she needed under the circumstances. There were some families, where, at mention of her name, she would have been received with formal politeness; but, she shrunk from an exposure of herself, and so, exhausted with long travel, and faint from heart-sickness, went wandering from street to street, in the long afternoon of a warm June day, until the burden of weariness was so great, that it seemed to her she must fall by the way. Many times, she more than half resolved to seek her father at his counting-room; but a dread of exposing herself there, held her back.

Four long hours of wandering in the street, and then, almost blind with headache, and scarcely able to stand from exhaustion, Lydia came a second time, to her father's door. It was now in the fast deepening twilight, and the day's warmth had given place to the evening's colder atmosphere, in which she shivered as if ague-stricken. Timidly she rung the bell — timidly, in sad consciousness that her right to enter was now a questioned right. She did not know the servant who opened the door.

"Is my father in?" she asked, making a movement to enter. But, quickly pressing the door against her, the servant said, almost roughly,

"You can't get in, Miss. Your father will not see you. So, don't come here again." And then shut the door in her face.

At this moment Adam came up the steps.

"Oh, Adam!" exclaimed the wretched creature, reaching her hands eagerly towards her brother. But he retreated from her as from a thing polluted and scorned.

"Don't touch me!" he said, roughly; and then hurrying past her, entered and closed the door.

"I must go in! I must see father!" murmured Lydia, rallying herself in desperation; and springing to the bell, she rang it vigorously. But no one came. Again she rang, but the door remained shut. Faint, and frightened at her alarming position — thus homeless and the night falling — Lydia stood for several minutes, leaning against the iron railing. Then, with slow, hesitating steps; halting and faltering; faint, and dim-sighted from pain and weakness, she moved aimlessly away, losing herself in dusky streets, down on which the darkness was rapidly falling.

CHAPTER XXIX.

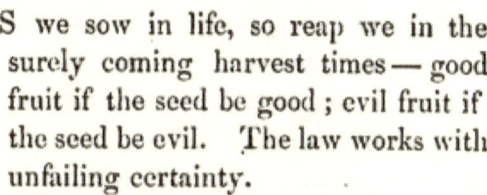AS we sow in life, so reap we in the surely coming harvest times — good fruit if the seed be good; evil fruit if the seed be evil. The law works with unfailing certainty.

True as this proved in the case of Adam Guy, it was also true in the case of Dr. Hofland. As the one sowed tares in his field, and found tares in midsummer and approaching autumn; so the other, having scattered wheat in his well prepared fields, gathered, in reaping time, full-eared sheaves of golden grain.

That one sharp experience in life proved quite effectual. Never again did the Doctor permit taste, ambition to make a good appearance, or a covetous desire for things beyond his ability to purchase, tempt him from the path of safety. Debt, after being once freed from its shackles, became an unknown element in his life; and warned by the memories of the past, he forced upon himself a well considered rule of expenditure, which always left him at the year's end a little better off than at its beginning. Thus, he remained free from anxieties and humiliating embarrassments, and so kept his mind above the depressing influences of care, that he could enter with full vigor into the spirit of his profession, and keep pace with its higher developments.

Some years before the period at which we have now arrived, the Doctor moved a second time from the humble home in which he had twice started in the world. In going up, on this occasion, the step was taken in all assurance that it was safe and right. His practice was largely on the increase, and he had a sum of money in bank considerably above the amount needed for extra furnishing. From his fall he arose again, wiser and warier. The discipline of a temporary misfortune, with its sharp humiliations and self-revealings, made him a truer, stronger and clearer seeing man. Out of the valley he came, and stopped not in ascending, until he stood far above the place from which he had fallen.

And sweet, also, had been to Lena, the uses of adversity. She had arisen, likewise, into a purer spiritual atmosphere, and now saw with clearer vision. And thus, ascending, she had drawn nearer to her husband. As his mind grew more and more in love with nobler things — as he grew wiser in the knowledge of those sublime truths which lead men up to an interior acknowledgement of God as the source of all life, she found increased pleasure in communing with his thoughts; and in the light of them, saw ravishing forms of spiritual beauty unrevealed to her own unaided vision. More and more were they growing into a oneness of life. He, the wise-seeing; and she, the wisdom-loving. Two minds were blending into one, in a sweet foretaste of eternal unity. Pleasant and instructive would it be to dwell with them for a brief season ; to look into their daily lives, and breathe in the tranquil atmosphere with which they were surrounded. But events bear us onward.

The Doctor was coming home from a visit in the neighborhood, just after dark, when, in passing a young woman, he noticed something in her face and manner that excited his interest. Her movements were slow and uncertain; and the look of exhaustion and almost despair that he saw on her countenance, as the light of a gas lamp revealed it to him for a moment, left on his feelings a most painful impression. The face was that of a child rather than of a woman.

"So young!" He sighed to himself, as he moved on, the thought of sin and shame crossing his mind, and sending to his heart a shade of sadness.

There was more than his usual tenderness in the manner of Doctor Hofland, as his lips touched the pure lips of his daughter Lena, on entering his home a few minutes afterwards.

"Must you go out again to-night, father?" said Lena, drawing her arm within his, after they had risen from the tea table, and holding him half playfully and half earnestly back from the hall into which he was about passing.

"Yes, dear; there are two or three patients who must be seen," answered the Doctor.

"Don't stay long," urged Lena. "It seems as if we never could have you in the evening."

"I'll be home in an hour."

"And that will be past nine o'clock," said Lena, with a shade of disappointment in her tones.

"We can make a long evening after that, if we will," was smilingly answered.

"I'm not so sure of that," returned Lena. "Ten to one, the office will be full of patients when you get back; or there'll be a call on the slate."

"In which case, dear, let us not only be thankful that the blessing of health is ours; but that, in God's providence, I have power to help the sick and suffering."

"I'm very selfish, I know," answered Lena, as she relaxed the firm hold with which she had grasped her father's arm; "but it is such a pleasure to have you at home in the evening."

"And such a pleasure to stay," replied her father. "Duty first, however." And taking his hat he went out. Doctor Hofland had only gone a short distance, when he noticed the same young person who had attracted his attention not long before. She was standing at one of the street corners, and seemed either awaiting some one, or to be in a state of indecision. As he passed, he drew near and made an effort to look into her face, but she started, with a timid air, and turning, walked slowly down an unfrequented street. The Doctor stood still and looked after her, feeling so much inclined to follow that he nearly yielded to the impulse. But, moving on his way again, he said, in his thought,

"Poor child! There is something wrong."

Scarcely satisfied with himself for letting an opportunity to help or save an unfortunate one, which Providence had placed in his way, pass unimproved, the Doctor walked onward, conscious of an unusual pressure on his feelings. Two visits were made, and he then crossed to a part of the city somewhat remote from that portion in which he lived. It was later than anticipated, when he returned to his own neighborhood, and he was walking with a quicker step than usual. Suddenly he stood still. A cry had fallen on his ears; a cry of terror — and the voice was a woman's. He looked in all directions, but could not determine from whence it came. Then it was repeated, and nearer, coming from an adjoining street, to which he hastened. At the corner, he met the same young woman who had twice before attracted his attention. She was running in a wild way. Seeing the Doctor, she fled to his side and caught hold of him, crying out,

"Oh, sir! Protect me, for heaven's sake!"

Two men followed quickly, pausing for a moment, as if to lay hold of her. But, on a closer view of Doctor Hofland, they moved away without speaking, crossed the street, and stood still on the other side.

The Doctor felt the girl's hand shaking on his arm, as they clung to him with a tight grip; and she pressed against him in a way that he understood to be from sudden prostration of strength.

"Who are you?" he asked, in a kind voice.

"Do you know Mr. Guy? — Adam Guy?" The choking voice that put the question trembled so that it was scarcely articulate.

The truth — or, at least a part of the truth flashed over the Doctor's mind in an instant. This was Lydia's child! He saw, in the dim light, her mother's old look in her face.

"Not Lydia Guy?" he said, in unveiled astonishment.

"O, sir, do you know my father?" And the wretched girl held on closer to his arm, and leaned still more heavily against him.

"What does this mean, Lydia? Why are *you* wandering alone in the street," asked the Doctor, assuming a serious tone.

"Are you Doctor Hofland?" said the girl, with a hopeful thrill in her voice.

"Yes, child. I am Doctor Hofland."

"O, sir! Wont you take me home for to-night! I've no place to go. My father is offended; and they wont let me see him. I came home from school to-day; but they wont let me in. I've been walking the street for hours. O, sir; for the love of heaven have mercy on me! Let me stay at your house to-night, and to-morrow I will go away. I'm not wicked, sir!"

'Poor child!" said Doctor Hofland, with a sob in his voice, as he drew her hand within his arm, "come home with me. For your mother's sake Mrs. Hofland will give you a mother's welcome." And with Lydia almost clinging to him, the Doctor moved on again.

"Home from school to-day?" asked the Doctor. "Did I so understand you?"

"Not exactly from school," Lydia answered, in evident embarrassment. "I left school nearly three weeks ago."

"To get married?" The truth suggested itself to the Doctor's mind.

"Yes, sir." Faintly.

"In opposition to your father's wishes?" said the Doctor.

"Yes, sir; or, at least without his knowledge."

"That was bad, bad, Lydia. I'm sorry."

She did not respond, and they kept on in silence.

"This is my house," said the Doctor, at length, as he paused. A painful sense of her humiliated condition was now so strongly felt by Lydia, that she drew back, murmuring —

"O, sir, I can't go in! Take me to my father's! He wont let me die in the street!"

"Have you seen him?" asked the Doctor.

"No, sir! My stepmother drove me from the house; and when I went back again, the servant refused to let me in."

Dr. Hofland reflected for some moments, as to what were best to be done.

"I think you had better remain with us to-night, and open your heart freely to Mrs. Hofland," said he, in answer. "She was your mother's friend, and will be a true friend to you. You are in no condition to walk farther."

Thus urged, Lydia yielded, and went in with the Doctor. Leaving her in one of the parlors, the kind-hearted physician sought for his wife, and, in a few hurried sentences, informed her of Lydia's presence in the house, and the cause thereof. Mrs. Hofland, the moment she clearly understood her husband, ran down stairs. Glancing hastily about the parlor, she saw Lydia in a half reclining position, on the sofa. Springing forward, she caught her in her arms, and thus prevented her falling forward on the floor. The intense strain of mind and body, from which she had suffered for hours, being removed, Lydia had sunk down quickly; and when Mrs. Hofland received her in her arms, every outward sense was locked.

"Better for her peace," said Dr. Hofland, as he stood looking upon the white face of Lydia, after she had been removed to a chamber, "were there no waking time for her in this world! Poor child!"

A slight convulsion moved over her face, even while he spoke.

"God knows what is best," he added, in a half regretful voice, as he recognized this sign of returning animation, "and He will temper the winds to the shorn lamb."

Then there were sighs, and low moanings and mutterings, as of one awaking from a sleep, which had been haunted by troubled dreams; then, the veil of unconsciousness was lifted, and she looked out upon life again — looked out, in surprise and tears.

Wronged and unhappy child! What a strange thrill pervaded her heart, as Mrs. Hofland drew her arm under her neck, and held her head lovingly against her bosom, kissing her, and weeping over her, for humanity's and her mother's sake. Love — tender, outgushing love, she had never known; and, as pity took the form of love, her heart swelled and pulsated with new emotions.

"Is this a dream?" she said, as thought grew clear, and she looked from the face of Mrs. Hofland, to that of her husband. "Where am I? What does it mean? Yes — yes! — I understand!" And, covering her face, she sobbed bitterly.

"Be calm, my dear," said Mrs. Hofland, affectionately. "You are with friends. Rest and sleep for to-night, are what you need." And she kissed her.

Lydia shut her eyes. How like she was to her mother! It seemed, to Lena's vision, that it was indeed the Lydia of her girlish days, who now lay so pale and still before her. Gently, and lovingly, did her hand pass over forehead and temples, smoothing back the damp hair, with soft, caressing tones.

"Oh, it is so hard," suddenly exclaimed Lydia, starting up in bed, the tears flowing over her cheeks. "I can't endure the thought, indeed, I can't!"

"What thought, my child?" asked Mrs. Hofland, striving, as she spoke, to press the unhappy creature back upon her pillow.

"The thought of being turned away from my father's door, as if I were one of the vilest! Left in the street, without friends or a home, to die, or meet a worse fate! Can you imagine a thing so cruel? It was not my father? No — no! hard as he may be, he is not so iron-hearted as that."

"To-morrow, we will talk of this, dear; not to-night," said Mrs. Hofland — "you are worn down, and excited. Rest and sleep are now demanded. And she gently bore her down again upon the bed. "Thank God, that friends unlooked for have been found — friends who will be true unto the last. I loved your mother very tenderly, and will love you, also, if you will lean upon me, and trust in me. Out of this bitter experience, God may lead you into unfailing pleasures. This may be only the beginning of a better and truer life. The way to mountain heights, is often first down into dark and gloomy valleys, out of which the soul comes weeping and trembling; but, God's angels were with it in the descent, its guides and comforters."

And, in such loving and true words, Mrs. Hofland won the confidence of Lydia, and soothed her into quiet. O'erwearied nature did the rest, locking her senses in sleep.

Morning found Mrs. Hofland early at Lydia's bedside.

"You are not well," she said, with undisguised concern, as she looked into her flushed face, and heavy eyes. "Your hands are quite hot," she added.

"My head aches badly," was the languid reply.

"Have you been awake long?"

"Yes, ma'am; a good while."

"The Doctor must see you," said Mrs. Hofland, turning away.

"Oh no, ma'am; I shall be better when I get up;" and Lydia made a movement to arise, but fell back, with a low moan.

"Nothing but what I expected," said Doctor Hofland, when his wife informed him of Lydia's condition. "It results from excessive fatigue, and mental excitement. She must be kept as quiet as possible."

"Will you see her father this morning?" asked Mrs. Hofland.

"He must be informed of his daughter's presence here; but, I have not yet decided whether to see him, or send him a note."

"It would be best to see him, I think," said Mrs. Hofland.

"I'm not sure in regard to that. A brief, rather unsatisfactory note, would set him to thinking; and, by the time he made his appearance here, if he should conclude to come, his mind would be in a better condition to hear all I might feel inclined to say, than if there had been no preparation for the interview."

"As you think best," said the Doctor's wife — "but, I have no faith in any reconciliation between Lydia and her family. If this marriage is, as I suspect, with a person in humble life, the indignation of Mr. Guy will know no bounds. If he is penniless, so to speak, the act will not be forgiven. The daughter may be suffered to come home, after a period of banishment, but her husband, never. So long as the sin of poverty stains his garments, he will be held off, as one despised and contemned."

"I fear as much," answered the Doctor; "but I can do no less than inform Mr. Guy of his daughter's presence in my house. Beyond that, the responsibility is with him."

CHAPTER XXX.

DID Mrs. Guy repent as the night came down? Did soft pity steal into her heart — pity for the unhappy child whom she had thrust so cruelly from her father's door? Were there no misgivings, nor relentings? Nothing of the kind. She had hardened her heart against Lydia long and long ago, and now only accepted the opportunity for pushing her aside with a resolute hand.

"That girl had the assurance to come here!" she said to Mr. Guy, in a tone which betrayed more than usual feeling. The children had retired and they were alone.

"What girl?" Mr. Guy started, and turned, in a disturbed way, toward his wife.

"Why, Lydia."

"Lydia!" The blood came darkly into his face. "Is *she* in town?"

"She was here to-day."

"What did she want?"

"To make it all up, I presume. To open the way for getting back here, with her beggar of a husband."

"By heavens, no!" exclaimed Mr. Guy. "No! No! No! Not while I have breath. What did you say to her?"

"I had nothing to say, beyond letting her understand that she had no longer any rights in this house."

"Did she ask for me?"

"Yes."

"What was your answer?"

"That you would be home in the evening."

"Has she called again?"

"I believe not." The woman lied outright. She knew that Lydia had been there a second time.

"Tell the servants not to admit her. Let her go to her husband. He owns her now. She is nothing to me, now!" The father spoke vehemently, being overcome with passion.

Mrs. Guy had already given that direction to the servants; but, as if acting under her husband's will, she left the room in pretence.

"It is done," she remarked, on coming back.

Mr. Guy did not respond. Fearful that he might relent, his wife said, in order to keep the balance of anger against Lydia —

"The coolness with which she came in was surprising; as if she had the same rights here as before."

"She will find out her error, I think," growled the father.

"So I imagine. The girl who does so wicked a thing, must be left to suffer the consequences."

Mr. Guy said, "I wish her no harm, but she must go the way she has chosen. I cast her off, utterly. I've said that already, and what I say I mean."

Satisfied that her husband was on the right side, Mrs. Guy did not press the subject too closely, lest, in simple opposition, she should throw in a word in favor of his child.

Sleep had become, of late, a chary attendant on the pillow of Adam Guy. If from any cause, he did not lose himself immediately on going to bed, and thought got free on the wings of truant fancy, he would lie half the night tossing about, and vainly seeking for oblivion; or, if, after getting to sleep, anything disturbed him, a like result followed. The morning found him exhausted as often as refreshed. From this cause, Mr. Guy was beginning to lose ground, as well physically as mentally. Neither body nor brain was sufficiently restored by the night's mission of health. The force of habit, in this thing, had begun to act with other causes, so that a tendency to wakefulness was steadily on the increase, and beginning to assume a grave aspect. It sometimes happened, that the whole night was spent in vain endeavors to lose himself, unconsciousness only being found as darkness gave way to the breaking dawn. Any disturbance was sure to be followed by a state of mental excitement precluding sleep.

It was not surprising, therefore, that, after retiring for the night, Mr. Guy found his "eyes set wide open," as he often expressed it, and that, instead of falling into a drowsy state, preluding sleep, his mind, in full activity, dwelt on the act and condition of his daughter Lydia. He was not so entirely lost to human feelings, that all touches of nature were dead. He could not wholly cast aside the memories of Lydia's sweet childhood; and now, it seemed, as if a hand were turning leaf after leaf, in the book of his life, and showing him records that stirred his heart with mingled pleasure and pain. As before intimated, there was a tender side for Lydia in the heart of Mr. Guy, indurated as that heart had become ; and, therefore, something more than anger afflicted him — and something more than the turbulence of bad passions drove sleep from his pillow on this night following her expulsion from his door.

Two or three times did Mr. Guy lose himself for just a moment — lose himself only to start up, wide awake, from some frightful dream, the action of which seemed extended over weeks. Then guided by reason, fancy would lead his thoughts to probable consequences that might have followed the turning off of Lydia. Where had she gone ? Should harm befall her, would not the sin lie at her father's door ? So strongly was this thought forced upon him, that he said aloud, defending himself from the assaults of accusing spirits,

"I did not do it !"

"Do what, Mr. Guy ?" His wife, from whose pillow busy thought had also banished sleep, startled by the words, arose, leaning on her arm, and bent over her husband.

"Turn Lydia from the house," answered the unhappy man, in a voice that made his wife's heart beat quicker and stronger, for it had a tone of pity and regret.

"Who did turn her from the house ?" she asked.

"You," was replied, in emphatic utterance.

"Not in obedience to my own will," said Mrs. Guy, in her usual cold and distinct utterance. "I acted only in your stead — did, just what I knew you expected me to do under the circumstances. Had I consulted my own feelings, there would have been a difference. So, don't, I pray you, lay any blame at my door."

Guy only responded to this by a groan, as he turned himself away, and shrinking down in bed, with his face buried in a pillow, tried to shut out thoughts that were troubling him beyond measure. The effort, however, was fruitless. His excited brain kept on with its morbid action, and gave new aspects to the relation in which he stood, not only to Lydia, but to his other children, and also to life separate from family interests. A cloud seemed to rest over everything — a blight seemed to have touched everything. Fear crept into his heart — fear of some impending calamity; the nature of which was undefined, — but, loss of money was involved, for riches in his estimation made the highest good.

So the night wore on, all the hour-strokes ringing on ears alert. Morning found Mr. Guy more than usually unrefreshed. Exhaustion had naturally succeeded. He was so nervous at breakfast time, that his hand shook as he raised his cup to his lips. Food, beyond a little coffee, he did not take; indeed, he was rarely able to eat anything in the morning.

Mrs. Guy watched her husband, half covertly, but with eyes that read every aspect presented. Two or three times she sought, by warily put questions, to lead him out, but was unable to get down to the current of his thoughts. While still at the table, a servant brought to Mr. Guy a letter, which had just been left at the door. He opened it in a hurried, rather disturbed way. Motionless and intent, his wife looked at him across the table. She saw a wave of feeling sweep over his face in a sudden impulse, as he glanced down at the signature before reading the letter.

"Who is it from," she asked.

But Mr. Guy did not answer. Partly averting his face, he read the communication.

"It is from Doctor Hofland," said Mr. Guy, rising, as he commenced refolding the letter. He spoke with an affected unconcern, that did not deceive his wife.

"What does he want?" she inquired, also affecting indifference.

It did not suit Mr. Guy to answer his wife, and so her question fell idly on the air. A moment, and the door shut behind him. If Mrs. Guy had obeyed the impulse that was on her, she would have followed her husband immediately from the room. But that would have been impolitic; showing too eager an interest in his state of mind as affecting Lydia. A brief delay, just for appearance sake, was only an act of prudence — brief as she made it, however, it proved too long, for in the interval her husband left the house.

On reaching his place of business, and retiring to his private counting-room, Mr. Guy re-opened and again read the letter of Doctor Hofland. It was as follows: —

"DEAR SIR — At a late hour last evening, I found your daughter Lydia in the street. She said that her stepmother had turned her from your door, and that she had been wandering, over the city, without food, for several hours. She was in a distressed and bewildered state. Two ruffians were in pursuit of her, at the time, and she was fleeing from them with cries of terror. I took her home, and she is now at my house, and I am sorry to say quite ill.

"Yours &c.,   EDWARD HOFLAND."

For over five minutes, Mr. Guy sat, pondering the answer he should make to this communication. Then taking up a pen, he wrote —

DOCTOR E. HOFLAND: — DEAR SIR, — For your kindness in protecting my disobedient daughter, I can do no less than give you my thanks. As she is at your house, pray render her all needed service. When she is well enough to leave the city, let her go to her husband. Send your bill of expenses to me, and it will be promptly settled. As for the girl, she must make her bed with the friends she has chosen. Her fault is one that will never be forgiven.

"Yours,   ADAM GUY."

Twice this was read over, and then torn up It was not in agreement with some interposing state of mind — whether of weakness towards his child, or regard for public opinion, cannot be said. After another period of reflection, he wrote again.

"DOCTOR HOFLAND: — DEAR SIR, — Your communication is just at hand. It has caused me acute pain. Do for my unhappy child whatever she needs, in common humanity, and hand me the charge. When well enough to be moved, send her back to her husband.

Let her understand that all attempts to return home will be fruitless. She will not be received. Even if I consented, all the rest would repel her. The disgrace she has brought upon the family is most keenly and indignantly felt.

     "Yours, &c. ADAM GUY."

 This letter shared the same fate. Another was then written.

 "DOCTOR HOFLAND:— DEAR SIR,— Your letter has pained me exceedingly. I was not at home when Lydia called; but, even if I had been, I should have declined seeing her. Of course, no one imagined that she was in the city alone, or without a lodging place. Where is her husband? Let her go to him as soon as she is well enough to leave your house. Please impress it on her mind, that all hope of a reconciliation with her family must be abandoned. Her fault is one that can neither be forgotten nor forgiven. From henceforth, she will be held as a stranger.

    "Yours, &c., ADAM GUY."

 This letter, no more satisfactory to the writer than had been the others, was sent to Doctor Hofland; but, in sending it, the thought of his child was not cast out from the father's mind. He might turn away from her — might shut his door against her — but, for all that, her image would creep in and haunt him with its perpetual presence.

 Notwithstanding his letter to Doctor Hofland was so worded as to close the door against all attempts at effecting a reconciliation, Mr. Guy, in his secret thoughts, held to the belief that the Doctor would not let the matter die. A hundred times during the day did his eyes glance towards the counting-room door, as he heard feet approaching, or the sound of a hand on the knob, certainly expecting to see the form of Doctor Hofland, or to receive a letter from him by the hand of a messenger. A higher respect for the Doctor, all at once took possession of his mind. For years he had held him in indifferent estimation, because he thought him poor and thriftless; and though he was now ranking high in his profession, and honorably spoken of by all, he lacked, still, in the eyes of the rich merchant, the money-sign of worth. But the fact that his daughter had found a refuge with the Doctor, and that the Doctor had addressed him a cold, formal note on the subject, in replying to which he had found it difficult to express himself satisfactorily, now conspired to work a change in his mind. He did not stand so far above the Doctor, nor hold him in such poor regard as before.

All day long, Mr. Guy looked for some response to his note, but none came. Night found him, with an unusual weight upon his feelings; and when he retired, it was to be haunted all through the weary hours by waking dreams, that found no pleasant changes. Not alone upon Lydia did his thoughts dwell. They went out upon the wide reaching sea, following after the boy whom he had committed to the waves. How would he be returned, if he came back at all? Better or worse? Alas! prophecy in the father's mind was dark, dark. In regard to Edwin, a warning word had already been received from the Principal of the school to which he had been sent. The boy's deportment was not good; and there was a hint indicative of something more serious. A proposition to bring him home had been met by a resolute objection on the part of his wife. Edwin was out of her way, and she did not mean to have him in it again — at least, not without a strong opposing effort. Then, recent losses in trade, and some large, current operations that began to look anything but promising, added to these causes of mental disturbance, and completely barred out the saving influences of sleep.

At midnight Mr. Guy was walking the floor of his chamber, unable, any longer, to lie in bed. For nearly an hour, he moved about, silent-footed, so that his wife might not awake, and then tried his pillow again. But, the brain was active as before, and went on creating, reviewing, and prophesying evil, with unabating fertility.

As day began to dawn, Mr. Guy, remembering that sleep had often come at this hour, resigned himself, in forced expectation of its stealthy approaches; but, the very state of mind thus induced, kept off the slumberous charms, and drove him from his bed an hour before the usual time of rising.

"You don't look well this morning," said his wife, regarding him with real concern. They were at the breakfast table.

Mr. Guy did not answer, though the remark produced a change in the expression of his face.

"Does your head ache?" Mr. Guy had reached up his hands, suddenly, and pressed them against his temples.

"No," — he answered, with an evasion of tone, as if

he did not care to be observed or questioned. "My head doesn't ache," and he withdrew his hands.

Mrs. Guy looked still more concerned, and, also, a little puzzled.

"Are you going to eat anything?" she asked, a little while afterwards.

"I've no appetite in the morning," he replied, pushing back his chair, and leaving the table, with his single cup of coffee not half emptied. His wife called after him, but he paid no heed to her remonstrance. Taking his hat from the stand in the hall, he went out.

Mr. Guy did not doubt but that he should find, among the morning's letters, one from Doctor Hofland. But, in this, he was mistaken — disappointed we might in truth say. The fact is, outspoken as had been his communication in regard to Lydia, a secret desire for mediation on the part of Doctor Hofland was felt. Not that he wished for a reconciliation with Lydia, or meant to let her consider him as other than a stranger — but his interest in his child was not dead; old, old chords of affection, twined in her earlier years, were pulling at his heart, and, while angry, he still desired to know how it was with her in the present, and how it would be in the future.

But, the day passed, and the curtain which his own hand had let fall between him and his daughter, was not lifted. Doctor Hofland had accepted his decision as final.

CHAPTER XXXI.

AS anticipated by Dr. Hofland, the illness of Lydia was temporary; the result of over excitement and fatigue. On the second day she was entirely free from fever, and able to sit up, though exhibiting signs of weakness. On the third day, she was strong enough to go out, when she made another effort to see her father; this time going to his store. There she learned that he was confined at home by a slight indisposition. Still fixed in her purpose to see him, she went to the house, and, as the door was opened, glided past the servant, so as to prevent its being shut in her face, as on a former occasion.

"Can I see Mr. Guy?" she asked, as she retired along the hall towards the entrance of one of the parlors.

"He is sick," replied the servant, "and cannot be seen."

"But, I must see him." Lydia's repressed excitement now manifested itself, and turning she ran up stairs, the girl who had opened the door following quickly, and calling out, "Mrs. Guy! Mrs. Guy!" in a half suppressed, warning voice.

At the head of the stairway, Lydia met her stepmother, who at once disputed her farther progress.

"I must see father!" exclaimed Lydia, attempting to pass; but a strong hand was laid on her shoulder.

"Didn't I say that she was not to come in!" demanded Mrs. Guy, in a low, fierce tone, addressing the servant, at the same time that she pushed Lydia back with a strength that the poor girl could not resist.

"Indeed, ma'am, and she slipped past me before I knew it was her," answered the servant.

"If you come here again, I'll send for a policeman," said Mrs. Guy, close to the ear of Lydia, and in a voice that chilled her like a sudden icy wind! "You don't belong to this house; so keep away, if you want to keep out of trouble."

And pressing steadily on to Lydia, she forced her down stairs and out into the street. Before the bewildered girl could recover herself from the sudden onset, the door shut heavily behind her. All this passed so rapidly, that Lydia scarcely realized the fact of a forcible expulsion from her father's house. But soon, an outflashing indignation fired her whole being, as her mind came up to a full comprehension of the outrage; and a wild spirit of revenge took possession of her soul. Her pale, stern face, startled Mrs. Hofland a little while afterwards, as she came into her presence hastily.

"Why, Lydia, child! Are you sick again? What has happened?" she asked, with much apparent concern.

"Not sick, ma'am, but outraged beyond all forgiveness!" Mrs. Hofland saw a gleam of fierce anger burn up in her eyes.

"How, child? How? Did you see your father?"

"No."

"Why not?"

"He was sick at home."

"Sick?"

"Yes: and instead of permitting me to see him, I was pushed down stairs and out of the house, as if I had been a thief!"

"By whom?" asked Mrs. Hofland.

"By a woman whom he calls wife! She put her hands upon me violently. I looked into her face and saw the tiger there — fierce and cruel. I know her now; and by all that I hold sacred in life, I will neither forget nor forgive her act to-day! From this hour, I am her implacable enemy." Her face, pale a little while before, grew dark with passion.

"Lydia! Lydia!" interposed Mrs. Hofland, almost frightened at the transformation which took place so suddenly. "Don't speak so."

"I'm in earnest, Mrs. Hofland," was answered, "madly in earnest! She has fully unveiled herself, and I know just what she is, and just what she means. I never liked her — never thought her my friend! — but the mask I so often tried to penetrate, has dropped from before her face, and I see her as she is. She never tried to win my love; and now I give her my undying hate!"

"Oh, Lydia! my dear young friend, hold back such thoughts. We must not speak of revenge, but forgiveness," said Mrs. Hofland, trying to calm the excited girl.

"It is too late, madam," answered Lydia, "I am only human. There are some things which cannot be forgiven, and this is one of them. As sure as I live, that woman shall see the day when suffering will come of this; suffering, if not repentance."

When Lydia went out on that morning, she was subdued in spirit; but now exhibited a fierceness that almost appalled Mrs. Hofland. She did not seem like the same individual. It was of no avail that she tried to soothe her feelings. The fire of passions burned on.

In the afternoon of that day, she announced her intention to return to her husband in the morning, asking for a sum of money sufficient to defray the expense, which would be refunded on her arrival at home. No opposition to this was made at the time; but, on the next morning, as she came down stairs, Mrs. Hofland met her with a sober face, and said,

"The Doctor thinks you had better remain a day longer."

"Why?" was the natural inquiry.

"You know that your father is sick."

"What of him?" A slight shade of alarm went over the face of Lydia.

"The Doctor was called in last night."

"To see father?"

"Yes."

"Is he so very ill?" Lydia's alarm increased.

"The Doctor has not said much in regard to your father; only he thinks you should not leave to-day."

Lydia sunk down upon the stairs, and became quite pale.

"Don't be frightened, my dear," said Mrs. Hofland, "there is nothing serious, I presume."

"But there must be, or else the Doctor wouldn't say anything about my remaining. And then, isn't it strange that he should be sent for? He's never attended in our family. Where is the doctor?" And Lydia arose quickly. "I must see him."

"He went out half an hour ago," replied Mrs. Hofland.

"To see father?"

Mrs. Hofland hesitated a little before answering this question, and then said —

"Yes, I think so."

"Was he sent for again?"

"No; but the Doctor said he would step around and see how he had passed the night."

"He must be very sick." And Lydia sat down again on the stairs, like one from whose limbs strength had departed.

"I did not infer from the Doctor's manner that your father was dangerously ill," replied Mrs. Hofland. "But, here he comes now."

The street door opened, and Doctor Hofland came in.

"Oh, Doctor!—how is father?" eagerly asked Lydia, starting to her feet, and leaning for support on the baluster.

The Doctor tried to speak and look cheerful, as he answered —

"He has not passed a very good night; though I found him quiet and easy." This increased rather than diminished the alarm of Lydia.

"What is the matter with him?" she asked, evincing the greatest anxiety.

"It is difficult to meet that question," replied Doctor Hofland. "The worst symptom is sleeplessness."

"Then, he has no bad sickness — nothing dangerous?" said Lydia, in a tone of relief.

"No, nothing that you would call dangerous." But there was something in the Doctor's manner that quickened anew Lydia's fears.

"Tell me what it is, Doctor. I ought to know." Lydia's voice was calmer and firmer. "You think I should not leave for home to-day. That, as I understand it, means something serious."

She fixed her gaze, searchingly, on Doctor Hofland, and waited for reply.

"Not necessarily very serious." The Doctor smiled in an assuring way. "As I said, the worst symptom is sleeplessness. Your father cannot sleep at night, and in consequence, his nervous system has become much exhausted."

"Isn't that strange?" asked Lydia, in a doubting, perplexed way.

"The condition is unusual," remarked the doctor.

"What is the cause?"

"I am unable to answer. The mind, I apprehend, however, has most to do with it. Your father has, all his life, permitted his thoughts to run too exclusively in a single direction. Some obstruction in the swift current has disturbed him."

Lydia dropped her eyes to the floor. She was not satisfied.

"I think we shall have a favorable change by to-morrow," said the Doctor. "But, you must not think of going home to-day."

"Do you regard the case as very serious?" asked Mrs. Hofland, when alone with her husband.

"I do," replied the Doctor. "Mr. Guy has not been able to sleep for or four five days, and that is bad — very bad."

"What is the cause?"

The Doctor shook his head.

"This trouble with Lydia has, no doubt, greatly disturbed him," said Mrs. Hofland.

"I think so."

"Has he mentioned her name?"

"No; but I can see that there is something on his mind about which he would like to speak with me."

"Didn't you try to lead him out?"

"Yes; but we were never alone. His wife hovers about him like a shadow. A number of times he sought to get her out of the room by asking for something, but she either rung the bell for a servant or supplied the want from resources at hand. I could see, plainly enough, that we were not to be left alone for a moment."

"She's an evil-minded woman, I'm afraid," said Mrs. Hofland.

"Cold-hearted, selfish, and designing. So she impresses me," replied the Doctor. "And it is clearly evident that Mr. Guy stands in fear of her."

"In fear!"

"Yes, fear is the word."

"I'm surprised at that. He never impressed me as a man over whom a woman could gain power," said Mrs. Hofland.

"That woman has power over him. Nothing is clearer. She moves about silently, almost stealthily; and has a low, smooth voice, over the modulations of which she holds perfect control. But, any one read beyond the first leaf in human nature, can see that she is deep and designing. Yesterday afternoon, on going with Doctor L—— to visit Mr. Guy, I noticed a gentleman leaving the house as we approached. Mrs. Guy came to the door with him; I observed that. On a nearer view I saw that it was Justin Larobe."

"The lawyer?"

"Yes. He was a student in Mrs. Guy's first husband's law office, and now, has a good practice at our bar. But, the man's reputation is bad. Cunning, specious, shrewd, and with fair talents, he has made headway in his profession; but I think him utterly void of honest principle. Why was he there? The question has come up a dozen times since I saw him leave the house. The answer that he had law business with Mr. Guy, and was there to consult him, does not satisfy me."

"What do you suspect?"

"I can't say that any definite suspicion has assumed shape in my mind; but I feel the shadow of something wrong."

After breakfast, and just as Doctor Hofland was preparing to go out, a messenger came from Mrs. Guy, saying that her husband wished to see him immediately. The Doctor stepped into his carriage and drove to the residence of Mr. Guy. In the hall, on his entrance, he met Mrs. Guy.

"Has there been a change for the worse?" he asked, seeing more trouble on the face of Mrs. Guy than he had yet observed there.

"He's acting very strangely, Doctor," she returned; "and insists on seeing you again this morning."

"How, strangely?" said the Doctor.

"Wildly, as if he were losing his mind. He frightened me dreadfully a little while ago."

The Doctor passed up stairs, hurriedly. He had been fearing decided symptoms of mental aberration. On reaching the door of Mr. Guy's chamber, he found it locked.

"Who's there?" called a voice from within.

"Dr. Hofland," he replied.

The key was turned, and the door opened just a little way.

"Come in, Doctor," said Mr. Guy, holding the door just far enough open for one person to crowd through. The instant Doctor Hofland was inside, the door was shut with a sudden movement, and the key turned.

Mrs. Guy knocked loudly for admittance, but her husband had withdrawn the key, and now held it tightly clutched in his hand.

"You can stay where you are, madam," he said, in a chuckling tone, and with a gleam of triumph on his face, that chilled the heart of Doctor Hofland, for both too clearly gave evidence of approaching insanity.

"I wanted to see you alone, Doctor," he remarked, a moment afterwards, the flash of light going out of his face, and his tone falling to one of grave earnest. "Sit down by the bed. I must lie down again. Isn't it strange that I get so weak, and nothing the matter with me — only just wakefulness?"

He threw himself on the bed from which he had arisen, and looked very earnestly at Doctor Hofland. He was about speaking, when some one rattled loudly at the door.

"Who's there?" he called, rising in bed.

"Me," answered a voice that was recognized as that of Mrs. Guy.

"Well, me can't come in!" he shouted, with angry vehemence." So just go off! I'm consulting the Doctor, and wont be disturbed." Then looking at Doctor Hofland, and lowering his voice, he said —

"I can't talk before her. She watches all my words so. You don't consider me very ill, Doctor, do you?"

"No, certainly not, Mr. Guy. This inability to sleep is unfortunate, however, and we must overcome it in some way."

"She thinks me dangerous." An expression of painful anxiety came into his face.

"Who?"

"Mrs. Guy."

"She has not said so to me."

"I'll tell you about it, Doctor." And the invalid leaned towards Doctor Hofland, and spoke in a hushed, confidential way. "She thinks I'm going to die."

"You imagine that," returned the Doctor, affecting a lightness of tone.

"No, sir; there's no imagination about it! It's just

as I say." Guy's manner grew excited. But, he changed in a moment, and speaking low and confidentially again, said —

"Doctor, we were old friends."

"Yes."

"Maybe it was my fault that we haven't always been friends." There was a conscious weakness in Guy's voice that touched the Doctor.

"We have not been enemies, even though friendly intimacy ceased," said Doctor Hofland, with kind encouragement in his tones. "And now, if I can serve you in any way, consider me as your best friend."

"Oh, thank you, Doctor: you're very kind." He spoke with animation. Then, as his voice fell to a sadder key, he continued. "She thinks I'm going to die; and, maybe I wrong her, but, in some cases you know, the wish is father to the thought."

A shiver ran along the Doctor's nerves.

"She's a deep woman," resumed Mr. Guy, seeing both surprise and incredulity in the Doctor's face. "You don't know her as I do. She only married me for my money. That, between us, Doctor. Of course, you'll not speak of it to any one. But, I want you to know it."

Guy sat upright in bed, and seemed debating whether to take the Doctor still farther into his confidence or not.

"I can trust you — yes — you're discreet; and besides you're an old friend," he went on, looking at the Doctor with a strange blending of weak confidence and solemnity. "She had a lawyer here yesterday!"

"Indeed!" The Doctor affected surprise.

"Yes. And what do you suppose she wanted me to do?"

"I'm sure I cannot tell."

"Make a will!"

"What?"

"Make a will!"

"Oh, that has been attended to long and long ago," said the Doctor, smiling. "You are not the man to neglect so important a thing."

"Of course I am not. But, don't you see what's in her mind?"

The Doctor did not answer.

"She wanted a will to suit herself, and then —"

A look of fear, almost horror, darkened the face of Mr. Guy.

"I'm sure you wrong her," said Doctor Hofland.

"I'm sure I do not. She's deep, deep, deep as the sea. You don't know her as I do. But, you don't think I'm going to die, Doctor?"

"Assuredly not."

Guy's face brightened.

"I've made my will," he said, leaning towards the Doctor, and laying a hand on his arm. "But, it wont just do. I must have a new will. Suppose I write one now. You'll be witness."

"Two witnesses are required, I believe," replied the Doctor, putting him off. "So it wouldn't stand if made. If you have a fair will, let it remain in force for the present. After you are able to go about again, a new one can be executed. That is my advice."

This seemed to pacify him, and he laid himself down in bed, breathing forth, as he did so, a long sigh.

"Shall I let your wife in now?" asked the Doctor.

"Oh, no! no!" quickly answered the sick man, starting up in bed again, and exhibiting a great deal of excitement. "I've got more to say; and she mustn't hear a word of it. What about Lydia?" This question was put abruptly, and with visible signs of pain.

"She is still at my house," replied the Doctor.

"Why don't she go to her precious husband?" His voice grew suddenly stern and angry.

"She was all ready to leave this morning."

"Well, why didn't she go?"

"I thought it best for her to remain a day or two longer."

"You did! For what reason?" Guy knit his brows, and looked suspiciously at the Doctor.

"She's been sick, and has not regained her full strength."

"Was that your only reason?" The eyes of the sick man still looked keenly into the Doctor's face.

"And you are sick."

"Ha! I thought so!" Guy jerked the words out sharply.

"Thought what?" The Doctor spoke very calmly, yet in a slight tone of surprise, not letting his eyes waver a moment in their return of the sick man's gaze.

"O, nothing," said Guy — "Nothing. It's just my foolishness." And he turned himself away, shrinking down in the bed. As he did so, the Doctor arose, and was crossing the room, for the purpose of unfastening the door, when, hearing the movement, he started up and called out, eagerly,

"No, no, no! Don't do that! Don't let her in! Come back here. I've a great deal more to say."

The Doctor returned to his patient.

"I got your letter about Lydia," said Guy, assuming a degree of calmness.

"And I received your answer."

"Well?"

"I did not think it the right answer, Mr. Guy. She is still your child — bone of your bone, and flesh of your flesh; and wrong doing on her part cannot alter the relation."

"The wrong is too great, sir.— The sin too deep. I have cast her off!" Guy's manner was stern.

"Not greater than our wrong doing and sin against God; but he never casts us off. Suppose He were as implacable as you are trying to be, what hope should we have in dying? Alas, death would then be full, indeed, of terrors! We might well tremble as it approaches. We must forgive, if we would be forgiven. As God's children, we ask mercy; but he has told us, that, in the measure we meet, it shall be measured to us again.— That if we do not forgive, we cannot be forgiven. Think of this, Mr. Guy."

He did think of it, as was plain from the fixed, solemn aspect of countenance, all at once assumed.

"*You* don't believe I will die?" he said, speaking half in confidence, half in fear; remembering the Doctor's assurances a little while before.

"With God are the issues of life," answered the Doctor. "We are in His hands, and He calls us in His own good time. It is for us to be always ready.

Some fall, suddenly, as you know, without warning; while others linger in wasting sickness. Of the day and the hour of his departure knows no man."

Mr. Guy's countenance became even more troubled. His eyes, that were very restless, glancing from one object to another, continually, had a staring look.

"Do you believe in a hell, Doctor?" he said, abruptly.

"Yes," replied the Doctor.

"Well, I don't, then! You can't frighten me with your fire and brimstone." He grew angry and excited. There was a wild play of all the facial muscles; and a fierce gleaming of the eyes. "So don't talk to me about hell and damnation! All stuff and nonsense!"

A hand rattled at the door. "Let me open it," said the Doctor. "It is your wife."

"No! No! *No!* I say no!" Guy shouted the last utterance of the word "No," with a mad vehemence. "She isn't going to come in here," he added, lowering his voice. "D'you know, Doctor," leaning closer and speaking in a hushed tone, "that I'm afraid of her. She's got designs on my life — I'm sure of it."

"Don't, I pray you, entertain so absurd an idea," answered the Doctor.

"There's one thing to be considered," said Guy brightening up, "She hasn't got the new will out of me yet; and I'm safe until that work is done. Ha! Ha! I'm ahead of her, Doctor — aint I?" He laughed in a low, chuckling way, that chilled the Doctor's blood.

"This being the case, there is no fear of any harm,

and so I will admit your wife," replied the Doctor, crossing to the door, against the sick man's remonstrance, and opening it. As Mrs. Guy, on whose face the Doctor read doubt and questioning suspicion, glided into the room, her husband shrunk down, and, turning his face to the wall, lay motionless as if asleep. Crossing to the bed, she spoke in a kind voice; but he made no response. Two or three times she addressed him; but he still refused to recognize her presence.

Doctor Hofland retired from the chamber, silently, motioning to Mrs. Guy, as he did so. She followed him out.

"It will not do to leave him alone, madam," said the Doctor.

"He's not growing violent, is he!" Mrs. Guy turned a little pale.

"His condition is such, that harm might come of his being left to himself. The door was locked when I came. Has that occurred before?"

"No."

"It must not occur again."

As the Doctor said this, Mrs. Guy, who stood near the chamber door, which was partly closed, moved forward with a spring. As she did so, it was shut with a quick jar, and the key turned on the inside. Her movement was too late.

The Doctor and Mrs. Guy looked at each other in surprise and alarm. Then the former caught at the door, and rattled it violently, calling, in a demanding tone —

"Mr. Guy! Mr. Guy! Open this door!"

"You can't come in; so, there's no use in making a noise!" answered Guy resolutely.

"If you will go down stairs, ma'am," said the Doctor, "I think he will open the door for me. Where is Adam?"

"At the store."

"Send for him immediately."

"I don't think that necessary, Doctor. And besides, Adam is only a boy, and has little influence with his father. If you can induce him to open the door, I will not leave the room again."

"I think you had better send for Adam, ma'am. We may need him. There is no telling how violent he may become."

"Then you really think he's losing his mind? O, dear!" And Mrs. Guy clasped her hands together, and put on a look of the deepest distress.

"We are losing time, ma'am," said the Doctor, in an anxious, impatient way. "Go down stairs, and send for Adam."

The Doctor waited until Mrs. Guy was out of hearing; then putting his mouth close to the chamber door, he called, in a loud whisper,

"Let me in, Mr Guy! She's gone down stairs."

But no answer came.

"Mr. Guy!" He repeated the call two or three times, but with no better success.

Listening intently, the Doctor heard the sick man moving about the floor. Then there was a noise like the opening of a window. A thrill of fear went through his heart.

"Mr. Guy!" He cried loudly, and struck the door two or three times with imperative raps.

"You can't come in," answered Guy, now speaking for the first time.

"Open the door! I've something very important to communicate."

"What is it?" The voice sounded nearer, to the Doctor's inexpressable relief.

"Let me in, quickly, before she comes up stairs."

"Ha, Doctor! I understand your trick. But it won't do. She's standing just behind you," was answered back.

"Upon my honor, no!" replied the Doctor.

"What do you want to say? Whisper it through the key hole;—I can hear."

"I shall do no such thing," replied the Doctor, assuming the tone of one slightly offended. "And I must say, that I'm surprised at your singular conduct. Is this the way for a gentleman to treat another, and in his own house? Open the door, or I shall go away immediately."

All was silent for several moments. Then the Doctor's quick ear detected a stealthy sound in the lock. The bolt sprung with a sudden click. He pushed, instantly, on the door, and it gave way to the pressure. Guy only permitted a small aperture to be made, and as the Doctor crowded through, shut the door and locked it again instantly.

"Now, sir, what have you of importance to communicate?" demanded Guy, turning upon the Doctor, with doubt and suspicion in his eyes.

"What am I to do with your daughter?" said Doctor Hofland, meeting the question, promptly, with one that involved important considerations.

This was unexpected on the part of Mr. Guy, as his momentarily suspended breath and blank countenance gave witness. He stood looking at the Doctor for a little while, in a confused way, as though he did not clearly understand him.

"What about my daughter?" he asked, at length.

"She is at my house. Your daughter Lydia."

"Oh!" A flash broke into his face, as we sometimes see it light up a cloud. "Lydia!" His voice was angry.

"Yes. You know she has been sick."

"Haven't I said my say about her, Doctor? Why do you annoy me further on that subject?" The manner of Guy was more subdued as he thus spoke.

"I have no wish to annoy you, sir," replied Doctor Hofland. "But you are her father."

"I'm not! I disown her!" He grew angry.

"Words are nothing against facts," said the Doctor, calmly. "You are her father. Think! I put it to your reason, and to your humanity."

This appeal staggered the invalid's weak brain.

"Now, what I wish to say," continued the Doctor, "is, that Lydia — your daughter — now at my house, does not possess the means of going to her husband. She has no money."

"Oh, well, if that's all, the remedy is plain. I'll send her some money. There was an air of relief about Mr. Guy, as he arose, and going to a wardrobe, opened it, and took from the pocket of his coat a pocket-book. "How much will she need?"

"Send her a good supply. If you will turn her from your door, don't let it be empty handed," said the Doctor.

Guy looked at him sharply for a moment, and then opening the pocket book, selected four five dollar bills.

"Will that do?" he asked, as he held them out.

"If it is best you can do, yes; but I think you should be more liberal. Remember, that she is to be no further cost to you — that her husband will support her now — and under this view, send her a liberal sum of money."

"You're a sharp one, Doctor!" A cunning smile flickered over the lips of Guy.

"Isn't there reason in what I say?"

"Maybe there is." And he commenced counting over the bank bills which had been removed from the pocket-book. "Fifty dollars. Shall I send all that?"

"Better make it five hundred," said the Doctor. "Haven't you a blank check in your pocket-book?"

"Five hundred dollars!" Guy looked confounded.

"She would have cost you twice that sum, probably, in a single year, but for her marriage," answered Doctor Hofland, with the utmost composure. "Don't you see?"

"That's so!" Guy's feeble mind was taken by this assault.

"Of course it's so! And as you are going to turn her out into the world — to cast her off from heart and home — don't let her depart in poverty as well as tears.

If you will not see the poor child, nor give her your blessing, send her enough to keep the gaunt wolf, hunger, from her door."

"I'll think about that," answered Guy, knitting his brows. "Here are fifty dollars for her, handing the bank bills to Doctor Hofland. "I never carry checks in my pocket-book, so I can't fill up one if I would."

"But you could give me an order on your firm to receive a certain sum for Lydia," urged the Doctor, not willing to give up the matter.

"Of course I could; but then, you see, I'm not going to do it. I'm too old a bird to be caught by your chaff, Doctor." And Guy smiled with an affected shrewdness, that was painful to witness. Then as a shadow came over his face, he asked —

"Hadn't you better give me a receipt for that money?"

The Doctor took a pencil and slip of paper from his pocket, and wrote a formal receipt, specifying that the fifty dollars were for Lydia.

"Will that answer?" he inquired, as he gave the receipt to Mr. Guy, who read it carefully.

"Yes, sir, that will do. I'm a man of business you must understand, Doctor. Never pay money without taking a receipt." And he folded the narrow bit of paper carefully, and laid it in his pocket book.

"I'll call round again towards evening," said Doctor Hofland, rising from the table at which he had pencilled the receipt. "I don't see that I can do anything for you now, unless you will permit the use of artificial means for inducing rest. And this I would most carnestly advise. It has already been delayed too long. Think, Mr. Guy; it is almost a week since you have had a refreshing sleep. Nature cannot stand this much longer."

"No — no — no! I've said no fifty times, Doctor! —haven't I?" responded Guy to this, showing much disturbance.

"And you may say 'No,' once too often," answered the Doctor, very gravely. "I see no help for you without an anodyne to quiet your nerves. Sleep is the only medicine that will reach your case, and I tell you, Mr. Guy, it must be had."

"I'll have the room darkened and kept very still. Sleep will come soon; I'm sure of it."

"A few light doses of morphia can do you no harm. Depend on it, my dear sir, nothing else will avail now." The Doctor's manner was very impressive. As he looked at Mr. Guy, he saw a glance of terror in his face. His eyes were directed to a remote part of the room. Suddenly he started up, partly raising his hands; a dead pallor overspreading his countenance. For a moment, his attitude was fixed; then as if some fearful apparition had faded from his vision, he caught his breath with signs of relief, and throwing himself back on a pillow, said, with considerable solemnity of manner,

"Doctor; I wish you'd say no more about morphia. The very name of it sends a shiver along my nerves. I've always had a horror of opium in any form — an idiosyncrasy, no doubt."

He closed his eyes and tried to keep very still; but, the Doctor noticed a constant twitching of the muscles in his face, accompanied by slightly spasmodic movements of the limbs.

"Very well, Mr. Guy," said the Doctor. "The responsibility must rest with you. Keep very quiet; have the room darkened; compose your thoughts. I will call again late in the afternoon; or, should you wish to see me at an earlier hour, send to my office. Good morning!"

As Doctor Hofland opened the chamber door, he quietly removed the key, and placed it in the hands of Mrs. Guy, whom he knew would be found on the other side.

"There is no hope for him," he whispered, "but in the administration of an anodyne. Unless sleep can be thus procured, wreck of mind, if not death, will be inevitable. Watch him carefully. A man should be in the room all the while. Have you sent for Adam?"

"Yes." But the woman spoke falsely.

"I will be here again, with Doctor L——, in the afternoon. Good morning!" And the Doctor retired.

CHAPTER XXXII.

OCTOR Hofland had dismissed his last office patient, and was preparing to go out for his afternoon visits, when a note was placed in his hands. It came from Mrs. Guy, and stated, that her husband having become violent, it had been found necessary to remove him to the hospital. This had been done, she said, at the instance of Doctor L——, their family physician.

Doctor Hofland read the note twice, and then, refolding it, with a grave, abstracted air, put it in his pocket, and left his office without communicating the fact to any one. The case being thus taken out of his hands was, of course, now beyond his reach; and the responsibility of looking after it removed. Except for the interest awakened in Lydia, he would not have been seriously affected by the event. A momentary throb of pain; a shadow of regret; a brief consideration of the case as involving a lesson in life — and it would have been, so far as he was concerned, as similar events in society, occurring on the outside of his personal relations. Except for Lydia, he would not have stepped aside to gain special information touching the removal of Mr. Guy;

but, as he would have to communicate the distressing fact on his return home, he felt under obligation to see Doctor L——, and learn from him the particulars involved. They were not satisfactory. Doctor L—— was scarcely as communicative as he could have desired, touching the condition of Mr. Guy at the time he was taken from the house. He had given the necessary certificate; but, only when questioned closely, did he admit the fact of not being present at the time of the removal.

"You do not know, then, whether violence had to be used?" said Doctor Hofland.

"There was no violence, I think," returned Doctor L——.

"How was his consent to the removal gained?"

"He was passive — indifferent — I believe. In a kind of stupor," replied Doctor L——, with an air of cool evasion that affected Doctor Hofland unpleasantly.

"In a stupor! Had he taken an anodyne?"

"Yes. O yes."

"I was not aware of that. Then you have seen him since morning?"

"I was there about two o'clock, and found him quite composed. Mrs. Guy said that he had consented to take a small dose of sulphate of morphia, the effects of which were plainly apparent. She then consulted me about his removal, to the hospital, and I thought it best to place him there while he was in a condition to be taken without resistance, and so gave a certificate, to be used if required."

Beyond this, Doctor Hofland could learn nothing.

After leaving Doctor L——, he thought of riding over to the hospital, which stood on elevated ground at the eastern end of the city, more than a mile distant, and seeing the resident physician; but the necessity of visiting a number of patients who required attention, prevented his doing so, and he returned home at nightfall, with no particulars of Mr. Guy's removal to communicate in answer to the eager questions which he knew would come from Lydia.

"How is father?" The words met him ere his foot was fairly beyond the threshold of his door.

The Doctor shook his head — looked sober — but did not answer. In what words should he convey the sad intelligence that must now be communicated?

"Is he worse, Doctor?" The pale, anxious face of Lydia grew ashen.

Doctor Hofland drew his arm around her, and leading her into one of the parlors, said, as he placed her on a sofa, and sat down by her side —

"Your father is better, I think, than when I saw him in the morning. An anodyne was administered this afternoon, under which he fell asleep. But, it was thought best by Doctor L——, to have him removed to the hospital, while unconscious through its influence."

"To the hospital, Doctor! Why to the hospital?" Lydia was wholly unprepared for the announcement which had been made.

"I should not have advised its being done, though his mind has wandered for the last day or two," replied the Doctor, in as even a voice as he could assume. "Sleep, under the anodyne which he has consented to take, will, I trust, restore the balance of reason."

The whole sad truth now flashed on Lydia. Her father was deranged, and in a hospital! Of little weight was the Doctor's last assuring sentence. She accepted the worst as true, and gave way to the most violent paroxysms of grief.

In the calm that followed, Doctor Hofland thought it best to communicate more particularly the state of her father's mind, and to prepare her for the worst, if it came. He had already learned enough about her husband, through her own admissions in regard to him, to feel seriously concerned for Lydia's future well-being and happiness. As far as he could see, the young man was little more than a social idler, who had sought to advance himself in the world by a rich marriage. At first, he thought of suggesting to send for him, in order that Lydia might remain longer in the city; but, after further consideration, it seemed not best to do so. On the following day, having ascertained that her father was in a better condition physically, though not mentally restored, Lydia concluded to return to her husband, Doctor Hofland promising to keep her informed of every material change in her father's condition. And so she departed, going out from the place of her birth a tearful exile — banished from her home — cast off — contemned — and with scarcely the feeblest hope of return. If it had not been for the stimulus of a keenly felt indignation and bitterness towards her step-mother, the wretched girl would scarcely have borne herself up. What had she to look forward to in life? That one act had separated her completely from all former conditions and associations, and she must now fall from luxurious ease and independence, where pride and self-love had been stimulated as plants in hot beds, down into obscurity and poverty — for that was the sphere of the husband she had chosen. The stern repulsion of her stepmother left no room for hope in that direction. She had clung, almost desperately, under the fear that appalled her spirit, after being denied admission to her father, to the belief that his forgiveness would be reached sooner or later; but all now was in danger of being lost. If this aberration of mind should become permanent, what hope of reconciliation with the family remained? Scarcely a shadow! Adam had already repulsed her in the cruelest manner: and, as for the rest, she had lived with them in perpetual strife, from the earliest times that she could remember. There was no love for her in any heart at home; and no one, therefore, to plead her cause.

For the week that followed, Doctor Hofland's engagements were more than usually pressing, and during that period he did not find opportunity for a visit to the hospital. On the ninth day after Mr. Guy's removal thither, he called on the resident physician. To his inquiry in regard to him, he received for answer, that Mr. Guy had been taken out of the institution three days before.

"Ah; I'm glad to learn that," said Doctor Hofland. "So the derangement was only temporary?"

"He was better, but not fully restored," replied the physician. "My advice was, to let him continue here for a longer period; but his wife came, in company with Doctor L——, and insisted on taking him home. I think, from what I saw in his face and manner, that he did not wish to accompany them. But, he made no resistance ; and as they assumed the responsibility of his removal, I, of course, could not object."

"How did he act, while here?" inquired Doctor Hofland.

"He was under the influence of morphia, when he arrived in company with his wife and Mr. Larobe."

"Mr. Larobe!" Doctor Hofland could not conceal the surprise he felt on hearing this.

"Yes, Mr. Larobe was with them. The effect of the anodyne did not pass off for nearly twelve hours, and we had fears, during a portion of the time, that the dose might have been too large. On becomeing fully awake, and conscious of his real position, Mr. Guy was shocked; but, after the first manifestations of surprise and indignation, he submitted passively ; though remaining silent and gloomy."

"Did he sleep again, without having resort to morphia?"

"Yes; but not for nearly twenty-four hours. He persistently refused to take another anodyne, and we did not care to use force unless as a last resort. Happily, nature did the work in her own way. Sleep came at length, with its salutary influences."

"Have you heard of him since he was taken away?" asked Doctor Hofland.

"No, but presume all is going on well."

"You think that he was decidedly better when removed?"

"Yes; I should say that he was better — though not

as well as I had hoped to see him become after natural sleep was restored. I'm afraid, should anything occur to disturb him seriously, that his brain will not be strong enough to bear the excitement."

"Did he seem clearly to realize the fact of having been placed in an asylum for the insane?"

"I think so."

"How do you judge as to the effect of this on his reason?"

"I think it would have been wisest on giving him the anodyne, to wait and see the condition of his mind after the effect subsided. The home surroundings and influences would have been more favorable to recovery than such as were met with here. At least this is my opinion."

"And one in which I fully agree with you," said Doctor Hofland. "Had I been consulted, as I should have been, I never would have advised the course that was taken. The case is a sad one, and I fear for the ultimate result. That intense, absorbing love of money, which seems to have been the ruling impulse of his life, often becomes a disease which you know to be as marked in its symptoms and progress as any laid down in the books, almost always terminating fatally to mind or body. Few men who thus abandon themselves to the one idea of making and hoarding money, live to what we call a good old age. The sword of their thought gets too sharp for the scabbard, and cuts its way through."

"Yes, that is the case in too large a number of instances. Mere money makers, if they survive either of the disasters you have referred to, are the feeblest and unhappiest of old men; self-tormentors, and inflictors of pain, or annoyance, on all who are so unfortunate as to be within the sphere of their influence."

"In this," remarked Doctor Hofland, "we have instructive illustration of man's folly in limiting the range of his thoughts and feelings to the little world of selfish interests — the poorest and meanest of which are involved in mere money getting, from the sordid love of money. Happiness is the end he sets in view — for that, all men sigh in present dissatisfaction and unrest — yet, how signally does the venture fail. Rich old men, who, from the beginning, set their hearts on mere possession, are almost always peevish, fretful, ill-natured, and dissatisfied with all around them. The exceptional instances are very few, and not highly creditable to human nature. If a man has nothing but money on which to subsist his spirit when he becomes old, he is poor and wretched indeed. Feebleness, or total loss of reason, comes, too often, as the mind's sad and only refuge from misery."

CHAPTER XXXIII.

AFTER writing, as he had promised, to Lydia, an account of her father's condition — stating that he was at home with his family again — Doctor Hofland dismissed the subject from his mind, as one not involving any special care or responsibility on his part, and heard nothing about Mr. Guy for several weeks. Then, with no little astonishment, he learned, that, when removed from the hospital, instead of being taken home, he was sent to a private asylum somewhere in the State of New York, and that, within a few days, a commission of lunacy had pronounced him hopelessly insane.

Not long afterwards it came to the Doctor's knowledge, that a guardian had been appointed for Mr. Guy's children, and his entire property removed from his control. As far as he could learn, Justin Larobe, the lawyer before mentioned, had been an active mover in the case, as legal adviser of Mrs. Guy, and was the duly appointed guardian.

"As well put sheep in the guardianship of a wolf," said the Doctor, to his wife, in communicating the information. "If anything could restore rational vigor to the mind of Adam Guy, it would be a knowledge of the fact, that his dearly loved treasures were in the grasp of this unscrupulous man. If there be any legal tricks by which the heirs can be defrauded, as surely as the sun shines are they doomed to poverty, even though their father's gold may now be counted by scores of thousands."

So covertly were all the proceedings growing out of Mr. Guy's mental state conducted, that Adam knew nothing about them until the decree establishing a guardianship was issued, and Mr. Larobe announced himself as standing to him in his father's place. Adam, now in his twenty-first year, could not repress his indignation.

"Why was I not consulted in this thing?" he demanded.

"You must put that question to your mother," was the lawyer's cool answer.

And, he did so, within the next ten minutes. The reply was characteristic of the woman, and significant of her purposes.

"Minors are not usually consulted in the matter of guardianships."

There was a cold sneer on her lips.

"In eight months, I will be of age, and then ——"

Adam checked himself.

"And then? Go on, sir."

"I will set aside this guardianship."

"Ah, will you?" The lady was cool and cynical. "Am glad to be advised of your intentions so early. Of course, your efforts will be successful, seeing that you are the youngest child."

Stung by her manner, and the cool defiance exhibited in her response, Adam lost control of himself, and indulged in a storm of invective, accusation, and threat, to all of which his step-mother listened without a sign of feeling. When he was done, she said, very calmly —

"Adam, there is one thing that I wish you to understand — you, and all others, claiming to make a part of this household. I am head and ruler; and my will, henceforth, is to be absolute law. Now, I am a peace-lover, and mean that peace shall be maintained here. I will have no more outbreaks of passion — no more 'scenes' — no more calling of hard names — no more fault-finding. If you, as your father's oldest son, are willing to remain on these conditions, well; if not, the world is wide enough for us all. Do you understand me?"

"Perhaps I do," answered the young man whose face had become deadly pale — pale from intense passion.

"Very well," said Mrs. Guy, coldly. She was about turning away, when he pronounced her name, sharply. She looked at him, with a glance of half indifferent inquiry on her face.

"I think I see your hand, madam."

There was the father's air of stern resolution in the boy.

"Ah?" The sneer in Mrs. Guy's tone, did not altogether conceal the sudden surprise occasioned by the words and manner of Adam.

"And I do not mean to be driven out, as you propose to yourself. I shall remain, and keep you under surveillance — you, and my precious guardian!"

"Adam! By——!" The subtle, self-poised woman, was thrown for an instant off her guard; but she caught up the lines of self-control the moment they dropped from her hands, and grasped them tightly again. In doing so, her teeth sank into her lip so sharply as to draw blood.

"By what? Go on." For a little while, the boy stood master of the position; but, only for a little while. His step-mother withdrew into herself again, and offered him no salient point of attack, thus baffling his courageous assault.

"I shall not repeat the admonition I gave you a little while ago," she said, with well assumed indifference. "Unless your conduct is in conformity with the rule I have announced as first in this household, you cannot remain; so, if the purpose to act as a spy is carried out you must put yourself on your good behavior — otherwise, the design will signally fail."

And, passing out from the room in which the interview occurred, she left Adam to his own thoughts, which were far from being as clear and determinate as he had sought to make his step-mother believe.

Only a few weeks were permitted to elapse from the date of Larobe's appointment as guardian, ere he gave formal notice to the firm in which Mr. Guy was senior partner, of his intention to withdraw the interest he represented; in other words, to dissolve the co-partnership, and change the status of property held in the business. Against this, Adam at once protested in the most resolute manner. He understood at a glance, the wrong involved in such a step — especially, the wrong to himself; for he had steadily looked forward to a position in the firm as partner; and, since his father's unfortunate loss of reason, to an actual representation of his interest.

"You will not agree to this?" he said to his father's partners, confident that they would interpose in some way to prevent so fatal a step from being taken — a step which must separate the estate, now held in trust for the heirs of his father, from large annual dividends in one of the most profitably conducted establishments in the city.

"We have no alternative," was the answer received by Adam. "Mr. Larobe is competent to order a dissolution, and we must submit."

"May I see the written agreement, under which the firm now exists?"

The partners looked at each other, inquiringly, hesitated, and then one made answer —

"That will be submitted to Mr. Larobe, as representative of your father's estate. He alone has a right to call for it."

Adam understood them now. Why should there be any hesitation about letting him see the agreement? He felt that there could only be one answer to the question. They were eager to seize the advantage offered, by which this whole business would fall into their hands; seize it at once, untramelled by any stipulations looking to an ultimate dissolution of the firm which might exist in the partnership papers.

And this was the truth. According to mutual agreement, expressed in writing, one year's notice of intended withdrawal from the firm had to be given. If this were adhered to, the interest of Mr. Guy could not be closed for a twelvemonth. But, Mrs. Guy, acting through Larobe, was eager to have all the property in a controllable shape as quickly as possible, and particularly before Adam reached the age of twenty-one. The business partners of Mr. Guy, accepting the opportunity for getting rid of their senior, by which they might grasp the entire establishment for themselves, were not unwilling to meet the views of his legal representative, and arrange for an immediate closing of his interest, which was done as speedily as possible.

It was in vain that Adam remonstrated, and insisted on seeing the articles of agreement; he only worked alienation towards himself in both parties, and gave a fair opportunity to his father's old associates in business to signify their wish to dispense with his presence at the desk he had been occupying for over a year. Removed in consequence from a position where he would have been able to keep himself advised in regard to the progressive withdrawal of his father's interest, with the amounts paid over, and the probable line of investments, Adam found himself completely baffled in his purpose to dog the steps of Larobe, who assumed towards him an impenetrable, half-offended reserve, on all occasions when they happened to meet. A small allowance of money was doled out through his step-mother, Larobe refusing to have any business intercourse with him, on the ground of having received an insult.

So completely had Mr. Guy separated himself from

social life — so entirely had he put confidence in money alone, as his best and most enduring friend, that now, in the great city where he had lived and grown rich, there was none to look after the interests of his children and protect them from wrong — none to examine into his unhappy case, and see that he was not held a prisoner on pretence of insanity, rather than in a salutary and needed confinement. Suddenly, a tempest had swept down upon the sea where he had spread his sails so long and proudly to the summer airs; and, though his vessel went down in the sight of hundreds, none were drawn to the rescue, and few, if any, were conscious of pity or sympathy. Having withdrawn himself from all community of interests — from all good-fellowship with his kind — ignoring, in the narrow spirit of mere " self-help," all the generous impulses of mutual help, there was none to care what might befall him in the voyage of life. And so, when disaster came, he was left to the help of his money-gods. If they could not save him, his case was hopeless. Alas! how hopeless it was proving!

A dead calm of months followed. John was still away at sea; but, letters from the captain of the vessel in which he had sailed as supercargo, gave a very discouraging account of his habits and conduct. He seemed to be completely demoralized. Lydia had made several attempts to effect a reconciliation with her family; but, all overtures were repulsed. The conduct of Edwin at school, was so bad, that the principal had written several times, threatening to dismiss him. In the midst of all this, the step-mother held herself at a cold distance from Adam and his youngest sister, Frances, who remained at home. Occasionally, Mr. Larobe came to see her, on business; but, these were rare occurrences, as she preferred seeing him at his own office, in order to blind Adam, who was always on the alert. While, so far as Adam knew, the intercourse between his step-mother and guardian was limited to rare interviews, not a week passed, without close conference between them.

One day — it was only a month or two from the time when the young man would reach his majority — Adam met Doctor Hofland. They had no acquaintance with each other. In fact, Doctor Hofland did not even know, by sight, the son of his early friend; but, hearing his name mentioned in a company, where both happened to be present, he drew him aside, and made inquiry about his father.

"No better," was the answer received.

"Where is he?" asked the Doctor.

"Somewhere in New York," replied Adam.

"In the city?"

"No, sir; I believe not; somewhere in the state."

"And don't you really know where he is?" The tone of surprise in which this was spoken, brought the blood to Adam's face.

"He's in an Asylum, near Troy." The young man stammered, and looked confused. Doctor Hofland was confounded; for, he understood this to be only a guess, or an evasion.

"If you are really in ignorance touching your father's condition, and place of confinement," he said, with considerable impressiveness of manner, "it is your duty to inform yourself as speedily as possible."

Doctor Hofland could not read, to his own satisfaction, the effect produced by this sentence. Adam was either shocked or offended. No answer was made; and the Doctor, feeling that he had no right to intrude farther, remarked on some current topic, and then left the young man to his own thoughts. He, soon after, missed him from the company.

On that same evening, and not very long after his brief interview with Doctor Hofland, Adam presented himself before his step-mother, and, with more agitation in his voice than he had the power to control, said abruptly, and with a significance of tone that startled Mrs. Guy —

"Where is father?"

"What do you mean? I don't get the drift of your question," said Mrs. Guy, so calmly as to conceal the quicker pulsations already leaping away from her heart.

"I simply said — where is father?"

"He's in an insane asylum. Were you never made aware of the fact?" How very even was her low-toned voice, in which was just apparent a vein of surprise.

"Of course, I'm aware of that fact; but, from some cause, the location has never been communicated. What my question involves, is the place of asylum."

"And don't you really know?" The expression of astonishment on the part of Mrs. Guy was very decided.

"That information you have, singularly enough, withheld."

"What do you mean, sir?" A flash sprung from the woman's cold eyes.

"Just what I have said, madam — that information you have, singularly enough, withheld. More than once, I have asked where my father was confined, but never received a satisfactory answer."

"Indeed! Well, you have shown yourself to be a loving and dutiful son!" How bitterly she sneered. "A year, almost, since your poor father was taken away, and yet, in all that time, you remain ignorant and indifferent about him — don't even know in what institution he is confined!"

"Will you now inform me?" said Adam, mastering, by a vigorous effort, the wave of passion, that was about sweeping him away, and revealing only a slight tremor in his voice.

"Certainly." Mrs. Guy smiled, and with a mock graciousness of manner that was excessively irritating.

"Where?"

"On Staten Island."

"In what asylum? Where is it located?"

"The institution is one of the best in the country," said Mrs. Guy, speaking with deliberation, and evidently seeking to gain time for thought. "We placed your father there, because we desired to secure for him in his unhappy condition, the wisest moral treatment, and the highest professional skill."

"What is the name of this institution?" inquired Adam.

"Woodville," answered Mrs. Guy.

"How is it reached?"

"Mr. Larobe can inform you. I have not been there."

"Although my father has been away from home for nearly a year!" Adam could not let the opportunity for a retaliatory thrust at his step-mother pass unimproved.

"His mental state is such as to render the presence of his friends unavailing for good. If that were not so, I should have been with him often," said Mrs. Guy, in imperturbable manner. "But I receive frequent reports of his condition, and have the calm satisfaction of knowing that all in human power to do for him, is done, and done under my direction. If you are in any doubt on this subject, I would advise an early visit to the institution; and, I must say, that your failure to do so up to this time, and general indifference touching your father, strikes me as very singular. Such indifference in a son, I have never before seen exhibited."

Adam was not skilled enough in human nature, to read the true meaning of all this. His step-mother was too deep for him.

"I shall not lie under that reproach long," returned the young man, angrily.

"I would not," was coldly answered.

And there the interview ended.

"Will you get for me, from Mr. Larobe, the exact locality of that aslyum?" said Adam, to his step-mother, on the next day.

"Why not get it from him yourself?" was replied, "I don't expect to see him very soon."

"Mr. Larobe and I are not on the best of terms; and it will not be agreeable for me to call at his office."

"Oh! I'm sorry. If I see him, I will ask him, of

course," said Mrs. Guy, with indifference. "But it is not at all likely, that he will be here for some time."

"Can't you send him a note?" inquired Adam.

"Yes, I could do so." Mrs. Guy's answer was not outspoken.

"Will you?"

"I'll think about it," and she retired from the room. Adam soon after left the house. It was beginning to shape itself more and more distinctly in his mind that something was wrong in respect to his father; and at last suspicion took the form of doubt in regard to his real insanity. Might he not be held in confinement, through the bribery of his keepers? The possibility of such a thing, once imagined, shocked the young man, and filled him with anxious alarm. After brooding over the suggestion for awhile, he determined to see Mr. Larobe himself, and learn all that he might feel disposed to communicate in regard to his father; and so, after conquering, with a strong effort, his unwillingness to meet the lawyer, he finally, under self-compulsion, entered his office.

"Can I see Mr. Larobe?" he asked of a young man who was writing at a table.

"He is engaged at present, but will be at leisure in a few minutes. Sit down;" and the young man pointed to a chair.

Adam took the chair. Adjoining the room in which he found himself, was another, the door of which stood ajar. In a little while, he noticed a murmur of voices coming from this room; and his ear soon detected, at intervals, the tones of a woman. Nearly a quarter of an hour elapsed, and still the murmur of voices went on. Adam grew impatient at length, and, rising, walked three or four times across the room.

"I'll drop in again," he said.

"He can't be occupied much longer," interposed the young man, who was a law student in the office.

Adam's hand was now on the door.

"I'll return in half an hour."

"What name shall I give?" asked the student.

"Say that Mr. Guy called."

"Mr. Guy! oh!" A gleam of intelligence lighted the young man's face. "Just wait a moment. I'll inform Mr. Larobe that you are here." And the student, first tapping at the door, pushed it open, and gliding into the back office, carefully shut the door behind him. He remained a few moments, and then returning, said —

"Can you call at four o'clock this afternoon? Mr. Larobe has several business engagements this morning, but will be pleased to see you at four."

"Very well. I'll endeavor to be here at the time you mention. Good morning." And Adam withdrew, feeling a sense of relief at having escaped meeting with the lawyer, towards whom he entertained a bitter animosity. Not long after his retirement, a lady emerged from the back office, and lingered in earnest conversation — speaking in low tones — with Mr. Larobe.

"I'll manage him; never fear," were the lawyer's last assuring words, as the lady, who was none other than Mrs. Guy, passed into the street.

Four o'clock came, and Mr. Larobe sat alone in his office, waiting for Adam Guy. But the young man did not make his appearance. His unwillingness to encounter the lawyer kept him from meeting the engagement. He preferred obtaining the information he sought, through the agency of his step-mother.

"Did you send a note to Mr. Larobe?" he asked, on finding an opportunity to be alone with Mrs. Guy in the evening.

"I did not," was coldly answered.

"You promised to do so?"

Mrs. Guy shook her head, at the same time that she compressed her lips firmly.

"You certainly did." Adam grew a little warm.

"I told you that I would think about it; and I have done so. From what passed between us last night and this morning, it is plain that certain base and inhuman suspicions in regard to me have entered your mind — suspicions that I feel as outrages. This being the case, I prefer not standing between you and Mr. Larobe, as the medium of intelligence touching your father. Go to him, and seek the information you desire."

Mrs. Guy showed unusual feeling for a woman of her cold temperament, and great self-command.

Adam was not prepared for this. His step-mother observed him closely; noting the effect of her opening assault, which was only preparatory of one of greater violence.

"You are to quick to imagine the supposition of wrong," he said, with a significant curl of his upper lip.

"What do you mean, sir!" demanded Mrs. Guy, with a fierceness of manner that startled Adam. He had never seen his step-mother so moved in his life — never felt such a fear of her as suddenly fell upon him.

"I said," he repeated, but with not half the firmness of his first utterance, "that you were quick to imagine the supposition of wrong."

"I am quick to feel the sting of a false and base insinuation, sir! — quick, as all true and honorable minds are," answered Mrs. Guy, with increasing indignation of manner. "And I tell you, sir, that you have gone just one step too far in a series of long continued outrages; and from this hour, I shall hold you at a distance. If you choose to place yourself in a position of antagonism, well; you have a right to the election, and also to the fruits thereof. Consider me from this time your enemy, if you will. I shall not shrink from the relation, depend upon it!"

Adam had in him too much of his father's dogged self-will to dream of stooping to conciliation.

"As you like," he simply said. Then added, with a threat in his voice, "The law is just; and I shall be of age in two months."

A gleam of cruel triumph lit the eyes of his stepmother; and she answered, in a hissing whisper —

"Those who take the sword, sometimes perish by the sword. Try the law, and abide by the law."

Both parties were too much excited to continue that wordy contest, as each felt; and so they mutually retired from the field. The quarrel was really of Mrs. Guy's seeking, though apparently brought on by Adam; but she was betraying herself a little too far under the pressure of feeling, and was glad to recede, lest some unwise utterance should fall from her lips.

CHAPTER XXXIV.

HAVING precipitated a quarrel with Adam, it was no part of his step-mother's programme to let the fires of antagonism go out for lack of fuel. Oil and faggot were always at hand, and furnished without stint, so that they were perpetually in a blaze. It ended as the step-mother desired. Adam left the house in a passion, vowing not to cross the threshold again, and took lodgings at one of the hotels. This occurred in a week after the brief interview mentioned as having taken place between Adam and Doctor Hofland. In order to make this separation complete, Mrs. Guy gathered everything belonging to the young man — clothing, books and other articles — and sent them to his address at the hotel, accompanied by a note couched in language that left him in no doubt of her purpose to hold him forever at a distance. He had turned from the door of his father's house, and now it was closed against him. He was out in the world alone, friendless and almost powerless. As he sat, in the small room at the hotel — so mean and poor compared with the one he left — sat there on the first night of his voluntary exile, and looked into the dark, mysterious future, his heart shivered, — his spirit grew faint and fearful. He realized more consciously than ever, that his step-mother was too strong and subtle for him, — that in her hands, he was weak as an infant.

As Adam sat thus, alone, looking with dismay at the prospect before him, Mr. Larobe and his step-mother were in close conference.

"You say," remarked Mrs. Guy, "that he has never called on you to ask the direction of the Asylum."

"I have not seen him," answered the lawyer.

"Just as I supposed it would be. Not a spark of filial love burns in one of their souls. I never saw such a heartless brood. If they had their father's money, he might be at the bottom of the ocean for all they cared."

"You think that Adam will not return here?" said Mr. Larobe.

"He can't return." Mrs. Guy's voice was soft and low, but it expressed the purpose of an iron will.

"I received a letter from Doctor Du Pontz to-day," remarked the lawyer.

"Well! What does he write?" The manner of Mrs. Guy changed instantly, and she leaned towards her companion with an expression of hope on her countenance.

"The Doctor wishes to see me, immediately."

"Ah! For what purpose?"

"Here is his letter." And Mr. Larobe read, — "I wish to see you at once, about Mr. Guy. His condition is less favorable than when I last wrote. Closer confinement, I fear, will be necessary; but, ere making a change, I think it best to ask an interview. Come with as little delay as possible."

"Is that all he says?" asked Mrs. Guy, taking a long breath.

"All." And Larobe handed her the letter, which she scanned closely.

Then they looked at each other, silently, for some moments.

"When will you go?"

"To-morrow. The case, as you see, admits of no delay."

"The Doctor writes with consummate prudence," said Mrs. Guy.

"He understands his business," replied the lawyer, with an expression of face that would have made innocence shudder.

On the evening of the third day following this interview, Mr. Larobe and Mrs. Guy met again.

"When did you arrive?" was the first, and natural question.

"Two hours ago."

"How is my husband?"

"That will inform you." And Mr. Larobe handed a letter from Doctor Du Pontz. Hastily opening it, Mrs. Guy read —

DEAR MADAM — I regret to inform you, that, within the last two weeks, all the symptoms in your honored husband's case have assumed more discouraging features. Mr. Larobe has visited him at my desire; and we agree, in the conclusion, that his safety will require a still greater restriction of liberty. The few gleams of intelligence which, lighting up now and then, gave us so much hope, seem to have died away forever, and left his mind in total darkness. It is exceedingly painful, my dear madam, to write in so disheartening a way of your excellent husband; but my duty is to state the case exactly. Have no fear that any harsh means will be adopted. These are not in my system. He shall be tenderly cared for, and permitted all the freedom consistent with safety. Very truly,

ALEXIS DU PONTZ, M. D."

"You saw him?" There was a look of inexpressable anxiety on the countenance of Mrs. Guy.

"Yes."

"He knew you?"

"Of course."

"What did he say?"

"He was violent — demanded a release — and threatened all manner of consequences."

Mrs. Guy's face grew pale.

"What did you say?"

"Nothing. It is not my custom to waste words on an insane man."

Larobe continued cool and self-possessed, but Mrs. Guy was nervous, and vaguely alarmed.

"Is there no danger of his escape?" she asked.

"He made several attempts recently," replied Larobe, "but the Doctor has now put him in such close confinement, that no apprehension need be entertained."

"I tremble at the bare imagination of such a thing," said Mrs. Guy, the paleness of her face remaining.

"You may fully depend on Doctor Du Pontz," were

Larobe's assuring words. "I have studied the man, and know him. Mr. Guy is safe."

"Adam will be of age in a few weeks." Mrs. Guy still spoke with manifest concern.

"Trust him with me, my dear madam. I understand his relation to the estate, and will take care that no disturbance of our plans originate with him." Thus the lawyer spoke in reassuring words.

"He will visit his father," said Mrs. Guy.

Larobe merely shrugged his shoulders.

"And see him " —

"No; that does not follow,"

"Doctor Du Pontz will hardly deny the son an interview. Were he to do so, suspicion would be aroused, and legal steps follow."

The lawyer bent close to the ear of Mrs. Guy, and whispered a brief sentence.

"Ah! I never thought of that!" she answered, light breaking into her face.

"Leave all with the Doctor," said Larobe. "He understands the case just as well as you or I, and will see that all things work to the good result we have in view."

Thus assured, the heart of Mrs. Guy took courage again.

Sooner than expected by Mrs. Guy, Adam's resolution to visit his father assumed the form of a present purpose. He called on Mr. Larobe within a few days after his separation from his step-mother, and asked for such information in regard to the Woodville Asylum as would enable him to find his way there by the directest course. Mr. Larobe treated him politely — even kindly — and not only gave him the information he sought, but also an open letter to Dr. Du Pontz, introducing him as the son of Mr. Guy, who wished to visit his father.

Adam left immediately for New York, and in an hour after his arrival was on his way to Staten Island. On reaching the landing, he hired a conveyance, according to the direction received from Mr. Larobe, and started for the Woodville Asylum, which was yet ten or twelve miles distant. A ride of two hours brought him in view of Doctor Du Pontz's establishment. Instead of a large and imposing modern edifice, as Adam had pictured it to himself, he found the Woodville Asylum to consist of an old-fashioned, two storied brick house, with high pitched roof and dormer windows, built, evidently, in the first years of the century by a well-conditioned farmer, or gentleman of wealth having rural tastes. The space covered by the main building and attachments was large; and the ground covered with fine old trees.

A chill crept along Adam's nerves as he passed from the avenue leading up to the house, and entered the grounds more immediately surrounding it. The old gate awry on the decaying posts, from which the paint had disappeared years before. The heavy box borders were ragged, broken, and untrimed; and the shrubbery, of which there was considerable, showed only partial and unskilled care. But, the walks were in good condition, and clean. A dead silence dwelt in the air — so dead, that to every footfall of the young man, an echo was stirred, and came distinctly to his sense of hearing.

As Adam ascended the steps leading to the door, he was met by a short, stout man, over fifty years of age, with a heavy black and grey beard, and a sallow countenance.

"Can I see Doctor Du Pontz?" he asked.

"That is my name," replied the short, stout man, with a slight French accent; bowing and smiling. "Walk in." And he moved back, giving way for Adam to enter.

The hall was spacious, having a broad stairway in the center. Doors opened into rooms on either side. From the hall Doctor Du Pontz conducted his visitor to a small apartment, evidently used as his private office, as it contained books, papers, medicine cases, and professional apparatus. On entering, Adam gave the Doctor his letter of introduction.

"Oh; Mr. Guy," said the Doctor, cordially, yet in a tone of sympathy, extending his hand and grasping that of Adam. "I'm gratified to receive a visit from you; though pained, of course, in view of the occasion."

"How is my father?" asked Adam, passing at once to the subject that was first in his thoughts.

The Doctor's face became serious.

"The case, I regret to say, is not as hopeful as could be desired."

"Can I see him?"

"O, yes." There was not a sign of hesitation. "But you must prepare yourself for a great change in his appearance. When mind gives way at the fearful rate witnessed in your father's case, bodily changes equally important, almost always attend the disaster. You will look upon a sadly altered man."

There was a tone of pity and sympathy in the Doctor's voice, that won a little on the confidence of Adam, and softened the unfavorable impression at first made.

"Do you think my father's case hopeless?" Adam's voice was husky and choked.

The Doctor gave a shrug, and arched his heavy eyebrows. Then, as his countenance fell back to its grave seriousness, he answered,

"Hopeless, I fear."

Adam caught his breath. The Doctor showed considerable feeling, and spoke kindly and sympathizingly.

"Wait here for a little while," he said, after conversing a short time with the young man; and Adam was left alone. In the ten minutes that passed before the Doctor's return, he did not observe with much care what was around him, for his mind was absorbed in the coming interview.

"Come," said Doctor Du Pontz, appearing at the door.

Adam arose and followed. So suddenly did the blood now flow back upon his heart, that it labored, half suffused, and felt like one on the eve of suffocation. All his soul shrunk from the meeting with his father about to occur. But it was too late to recede.

"Will he know me?" the young man found voice to inquire, in a suppressed whisper, as they paused before entering a room on the second floor of one of the additions which had been made to the main building — an addition not seen in approaching the house.

"I cannot say," replied the Doctor. "He is not much inclined to notice any one. But, he may remember you."

And the Doctor turned the key, that was in the lock on the outside of the door, and pushing it gently open, passed in, followed by Adam.

Sitting near the window was a man, to all appearance, sixty years of age, his face covered with a short, grizzly beard. He did not stir, nor seem in any way surprised or disconcerted by the intrusion; but fixed his wild looking eyes intently on Adam, who, recognizing scarcely a feature, advanced quickly, and holding out his hand, pronounced the word "Father!" in an eager tone.

The man started, and bending towards Adam, who had already grasped his hand, looked at him curiously, yet in evident doubt.

"I am Adam, father; your son Adam," said the young man, with much tenderness and feeling.

"Adam? Adam? My son Adam? I thought he died a long while ago." The man looked doubtfully at Adam, and then in a mute questioning way from him to the Doctor.

"O, no," said Doctor Du Pontz, falling in with his humor, "that was a mistake. He didn't die. He was dangerously sick for a long time, and word came that his illness had terminated fatally. But, you'll remember I told you last week, that this was true, and that he would be here in a few days."

"He looks like my boy." And a light shone in the vacant face, as this was said.

"I *am* your boy, your *own* boy, father!" Adam's voice shook, and his eyes were blinded with tears.

"Are you indeed? Well, I couldn't have believed it. They say the dead come to life again, sometimes."

And the poor old man smoothed, with both of his hands, the temples and face of Adam, along whose nerves the cold, unearthly touches sent a shudder.

"Can't I go home with you, my son?" he asked, in a plaintive, pleading voice. "I don't like to stay here. I want to go home."

"Why don't you like to stay here?" asked Adam.

Instead of answering, the old man threw a half fearful look at Doctor Du Pontz. Adam turned quickly, and saw an intimidating glance in the Doctor's eyes.

"Why don't you like to stay?" Doctor Du Pontz repeated Adam's question, and in a tone that, to all appearance, invited confidence.

But, no answer was returned. That one glance from the Doctor seemed to have touched him like a spell. His face lost all signs of intelligence or feeling. He receded from an awakened state of dim, hopeful consciousness, to the gloomy caverns where his soul had been dwelling. Fruitless were Adam's efforts to call him back again. Voice and words failed to penetrate the region of conscious life. He sat, still as a statue, with an unchanging countenance, and eyes that never, for an instant, lifted themselves from the floor.

"I am going, father," said Adam, after exhausting all the means of gaining attention that were suggested to his mind.

No response was made. Adam partly turned, and moved a step or two in the direction of the door; then stopped and waited; but there was no recognition of the movement.

"Good-by, father!" Adam's voice trembled. He

came back a few paces, and held out his hand. "Good-by, father," he repeated. But, the form before him remained immovable. He stooped, and lifting one of the impassive hands, said —

"I'm going now father."

The touch aroused the old man. Springing to his feet, he caught the shoulder of Adam with a strong grip, and holding him off at arm's length, glared wildly into his face. Instantly Doctor Du Pontz was between them, and, by the exertion of great strength, broke away the hold on Adam, and pressing back the now foaming and raving maniac, called, in a quick, warning voice for the young man to leave the room instantly, an injunction which he did not fail to obey. On the outside of the door Adam stood, all in a tremor, listening to the struggle that still went on, and which continued for nearly a minute. Then the clanking of a chain chilled his blood, and with a sickening heart he made his way to the office in which he had been at first received, there to await the Doctor's return.

Soon Doctor Du Pontz joined him, his dress in considerable disorder, and his sallow face flushed.

"Poor man! You see how sad the case is. Mind nearly gone." The Doctor spoke in a tone of pity.

"Are such dreadful paroxysms frequent?" asked Adam.

"No. If all causes of excitement are avoided, he remains calm and indifferent," replied the Doctor.

"He knew me," said Adam.

"There was, certainly, a partial recognition, and the disturbance which followed was a consequence. We are obliged to keep all exciting influences as far away as possible."

"Do you see any improvement, taking his condition now, and comparing it with his condition one, two or three months back?" inquired Adam.

The Doctor shook his head.

"Is there any change?"

"I think so."

"Unfavorable?"

"Yes."

Adam sighed heavily, and remained silent.

"It pains me," said Doctor Du Pontz, "to be obliged to speak with so little encouragement; but truth compels me to affirm, that I see little in your father's case to inspire hope. But, of one thing you may be assured, while in this establishment, he will receive the kindest care, and the best moral discipline his case demands. Time, and a wise patience, may bring salutary results. As in diseases of the body, so we say in diseases of the mind, while there is life there is hope."

Adam went back from Woodville, sadder than when he came, and with a darker cloud resting on all his future. He felt a sense of weakness creeping into his soul, as if forces, impossible to be conquered, were arraying themselves against him. Between his step-mother and Mr. Larobe, evidently existed a league; and there was little doubt in Adam's mind as to the object. But, what could he do to thwart the evil purpose they had, as he believed, in view? Nothing, certainly, until he attained his majority.

"A few weeks more, and then!" So Adam said to

himself, many times, as he journeyed back. "A few weeks, and then this guardianship, so far as I am concerned, must cease."

But what then? The answers were far from clear. He would take counsel, and demand a legal adjustment of his relation to his father's estate. The law would put him right! But Adam did not know the law.

15

CHAPTER XXXV.

 A few weeks — and they had nearly expired. In ten days, Adam Guy would be twenty-one. He had already taken legal advice, and was preparing to put his step-mother and Mr. Larobe on the defensive in regard to his father's estate. On the very day of reaching his majority, a note from his counsel was to signify his will in the case. All the assurances he received were of the most emphatic character. He was told that the Orphan's Court would order a division to him of so much of his father's property as, in heirship, he was entitled to receive. Beyond that, he had no concern. If his brother's and sister's portions were alienated or squandered, under the guardianship, it was of little concern to Adam. He was for himself, and for no one else. Already he stood separated from them; and after getting his share of his father's property, he meant that the alienation should be complete. They must not become clogs or hindrances to him on his way upward!

Such were Adam's thoughts and conclusions as he sat alone in his room just ten days before the limitations of minorship were to be removed. There was a knock at his door.

"Come in."

A servant entered and handed the young man a card. It bore the name of JUSTIN LAROBE.

"Show him up," said Adam.

A few minutes were passed in wondering suspense, not untouched by anxiety.

"What does *he* want?" more than once found an almost audible utterance.

Hastening feet were soon heard, and then as the door swung open again, Adam arose quickly, with a half uttered exclamation on his lips, and a look of alarm on his face. Larobe confronted him with a pale, agitated countenance.

"Oh, Adam!" exclaimed the lawyer, speaking in a tone of anguish. "Such a dreadful thing has happened!"

"What?" asked the young man, with a look of terror.

"Your father."

"What of him, Mr. Larobe?"

"Is dead!"

"Dead? Dead!"

Mr. Larobe's hand shook as he drew a letter from his pocket and handed it to Adam. It was from Doctor Du Pontz, briefly conveying information that Mr. Guy had escaped from his room in the night and fallen from a window. In the morning he was found, lying on his face, which was cut and bruised by the fall, dead and stiff — life having been extinct for some hours.

"My poor father!" sobbed Adam, hiding his face with his hands, as some waves of natural emotion swept over his heart. A short space of time the lawyer stood silent, until the first outburst of feeling had subsided. Then he said,

"I have not yet seen your mother; and dread being the medium of such horrible news. How shall I break to her this appalling intelligence? Will you go with me?"

But Adam shook his head.

"Then I must see her alone," said Mr. Larobe, with visibly regained composure of mind. "The body must of course be brought home. I will make such arrangements here as the case requires, and then go on in the evening train and reach Woodville to-morrow. In the mean time, go to your mother, and give her all the consolation in your power. Let this sad event obliterate all unkindness. I will write a few lines to Edwin and Lydia.

The paleness and agitation had already departed from the lawyers face. He was composed and business-like in his manner.

Adam was too much stunned by the intelligence of his father's death to offer any reply, and Mr. Larobe, having discharged his duty in making the announcement, hurried away to the step-mother's residence, whither Adam followed in the course of an hour. Mrs. Guy was overwhelmed by the dreadful intelligence, Adam found her bathed in tears, and overflowing with grief's most eloquent language. She blamed herself for having ever consented to the removal of her husband to an Asylum.

"I could have guarded him with sleepless watchful

ness," she said, " and so prevented this terrible calamity."

But, the son was not deceived. In acting a part, Mrs. Guy, like most actors, gave the lie to nature. Adam looked on in silence, contempt, and suspicion.

After a brief interview with his step-mother, who soon regained her usual calm exterior, Adam retired from the house and went to the office of his lawyer in order to state the fact of his father's death. Sorrow made no part of the concern that now weighed upon his mind.

" Is there a will ? " was the lawyer's first inquiry.

Adam could not answer the question.

" All depends on that. If there is a will, legally executed, its provisions bind the estate ; if there is no will, the law of inheritance comes in to divide the property. We must wait and see. Your father was too careful and systematic a man to neglect so important a thing as his will."

Anxious and impatient for the time when all suspense would be removed, Adam passed the next few days in a state of restless uncertainty. Mr. Larobe returned, in due course, with the body, preserved in ice ; but the face was so blackened and disfigured that not a feature was recognizable. By previous arrangement, the funeral took place on the day following. Of the severed and alienated house-hold, all were present but John, who was still at sea. Lydia was scarcely recognized by Adam or her step-mother. No one spoke to, or noticed her husband, after the first cold introduction. Although Lydia arrived on the day previous to that designated for the funeral, she was not invited to remain, and after sitting for an hour or two in the heart-chilling atmosphere of her old home, retired with her husband.

There were but few in attendance on the next day; and of sincere mourners, perhaps not one. No hearts — not even those of Mr. Guy's children — garnered sweet memories of the departed, or wept for a loss felt to be irreparable. The earth, as it fell heavily on that coffin lid, was not heaped on one whose spirit had linked itself with human loves and human sympathies; and no broken bonds or rent tendrils bled or quivered in next to mortal anguish with the pain of separation. Tearless eyes had looked upon the coffin as it descended from view, and tearless eyes turned from the spot where it disappeared — a spot unconsecrated by sorrow, and never to be visited with any loving interest.

Is the life worth living, that closes in such a death and burial? And the beyond? Thought comes back, shuddering, from the beyond, and we ask, " What is the man's state now?"

One act more. Word was passed, quietly, to Adam and Lydia, that, immediately after the funeral ceremonies, their father's will, found among his papers, would be read. This was the first advice received by them touching the existence of a will. They, therefore, returned to the house, and met the assembled family. Mr. Larobe was in attendance, sitting at a table in one of the parlors, with the will lying before him. Formally he broke the seal, in presence of all, and in the death-like stillness that followed, read the document in a calm, distinct voice.

After the usual brief preliminary matter, the children

of the testator were named in order, with the sum each was to receive from the estate. Adam's share was twenty thousand dollars. John's ten thousand; but this was in trust; he to receive only the annually accruing interest. At his death, the principal would pass to the residuary legatee. Lydia was next mentioned. Her portion was only one thousand dollars! Edwin and Francis were to receive ten thousand each, and a like sum was willed to each of Mr. Guy's three children by his second wife. All the residue of the estate, real and personal, was bequeathed to his " loving wife, Jane," who, jointly with Justin Larobe, were constituted guardians to the children not yet of legal age. Larobe was named as executor to the estate.

No sentence of approval or blame appeared any where in the carefully worded, and strictly legal document, nor were any reasons for a bequest given. The only apparent sign of human feeling, was in the words " my loving wife, Jane."

Several minutes passed, after the lawyer had finished, before the succeeding silence was broken. The first movement was on the part of Lydia and her husband, who arose, and retired from the house, without the utterance of a word. Adam next withdrew, and in silence, also departing from the house. Edwin sat for a little while stunned to bewilderment by the announcement of his small share in an estate which he had complacently estimated at the value of several hundreds of thousands of dollars, and then retired also, going up to his room, for he yet claimed some right to a place in his father's house.

Adam's steps were directed to the office of his counsel and he went thither with hurrying feet.

"There is a will!" he said, with strong excitement in his manner on entering.

"Has it been read?"

"Yes."

"What are the provisions?"

"Oh, horrible! Scandalous! That woman gets nearly the whole of my father's large property. I shall contest the will."

"What is the date?"

"I didn't observe," said Adam.

"That is important. If executed since the aberration of mind which preceded his death, it can be set aside. As soon as we have the probate, I will look to the date. But, what is your share under this will?"

"Twenty thousand dollars." Adam tossed his head in contempt of the paltry sum.

"And what of the other children?"

"Only ten thousand each, except in the case of my sister Lydia, who threw herself away in a beggarly marriage. She is cut off with a single thousand."

"How much do these sums amount to in the aggregate?" inquired the lawyer.

"Six, at ten thousand each, make sixty thousand dollars. My portion of twenty thousand, and Lydia's portion of one thousand, added, would make the sum of eighty-one thousand dollars."

"How large is the estate?"

"Roughly estimated, say two hundred and fifty thousand dollars."

"Of which your mother comes in for nearly one hundred and twenty thousand."

"Yes — she a mere interloper — a woman who only married my father for his money — she to get the lion's share! My blood boils in my veins!"

"Let us see how the case stands, should this will be defective," said the lawyer. "Your mother's one-third of the whole estate, valued, we will assume, at two hundred and fifty thousand dollars, would be something over eighty-three thousand. There would remain, say one hundred and sixty-six thousand dollars, to be divided among — how many children?"

"Eight," replied Adam.

"Eight into one hundred and sixty-six, a little over twenty times. I don't see that you would be any better off, Mr Guy — you, individually, I mean."

Adam's countenance fell.

"As the will now stands, *you* are to receive all the law would give, had your father died intestate. The hard features of the case lie with your brothers and sisters."

"I see — I see." Adam's face had grown deadly pale. Only twenty thousand dollars as his share in an estate, of which already greedy eyes had appropriated more than one half. Just how this appropriation was to be made, had not been settled in the young man's mind; but, he had cherished a vague impression, that, after attaining his majority, the estate would come under his control, and be left almost entirely to his management. In that event, no delicate scruples as to others' rights, when set against his strong love of money, would hinder the execution of his will. Justice, humanity, integrity, except as safe virtues, were not involved in his rules of action.

"The rights of your brothers and sisters are to be considered," the lawyer said. "As the will now stands, they will only receive sixty-one thousand dollars. If broken, one hundred and forty thousand will be divided between them."

"But *I* will get no more!" said Adam, sternly.

"You will get no more," replied the lawyer.

"Then I shall not move a finger in the case. Stand the brunt, and expense, and delay of a law-suit, with no prospect of a dollar's advantage? Adam Guy is not so great a fool! Let them fight who have something to gain. If twenty thousand is all I am to receive, I will take the paltry sum and make the best of it."

And saying this, Adam withdrew. He was satisfied, that the will which his father had made did not express his true purpose; that it had been extorted from him in some moment of weakness, or derangement; and that, if an attempt were made to resist its provisions, the court would pronounce it null and void. But, why should he involve himself in the cost, vexations, and delays of a law-suit, which, if decided in his favor, would leave him no better off? Adam was shrewdly selfish enough to comprehend the folly of such a course, as affecting his own position, and no motive of good will, or care for his brothers and sisters, could induce him to enter a contest designed alone for their benefit. The sentiment, "every one for himself," expressed accurately his state of feeling. He comprehended no interests but his own.

The more carefully and soberly Adam considered the

stipulations of his father's will, the clearer did it become manifest, that his wisest course was to accept its provisions as affecting himself. There were too many to share in the results of a law-suit, as against the instrument, if carried to a successful issue. And besides, if he contested the will, he must forego its benefits, and so be kept from any share in the estate until a final decision by the last court of appeal to which the case could be taken.

In considering the case of John and Lydia, Adam had no fault to find with his father's will. Ten thousand dollars in trust for John, he considered a fair appropriation to one of his spendthrift habits. The devise in trust, so that only the interest could be used, met his entire approval. As for Lydia, one thousand or one hundred was all the same to him. They had parted company in life; their roads had taken a sharp divergence, and could never run side by side again. Towards Edwin and Francis, Adam was coldly indifferent. He could part company with them also, and not suffer a pang. As for his half brothers and sister, they not only shared the dislike with which he had always regarded his step-mother, but were held to be interlopers — intruders, who had come in to wrong, and who had wronged by their presence the first heirs of his father. If we put the case stronger, and say that he hated them, our words would more accurately express the truth.

So, without a movement looking towards investigation, although he entertained the strongest suspicions in regard to the means by which this will was obtained, Adam, on reaching his twenty-first year, accepted his share of the estate, which was promptly paid by the executor, and then, resolutely, in heart, turned himself away from all kith and kin, resolved to be alone in the world, and all for himself.

Three months afterwards, John returned from his sea voyage. He was changed, and for the worse. From being sensual and depraved he had become cruel and desperate also. During a portion of the voyage homeward the captain had been obliged to put him in irons for mutinous conduct. He exhibited no natural emotions on hearing of his father's death; but asked, almost immediately, in regard to the will. His first interview with Mr. Larobe, during which a copy of the will was placed in his hands, took on a stormy character. He denounced the instrument as a fraud, and swore that he would contest it through every court in the land. From the lawyer's office he went in search of Adam and soon, by his violent language and unjust insinuations, stirred his brother's cooler blood with passion. Sharp words passed and anger grew hot. Both were mad and blind. In his ungovernable rage, John struck a blow that scarcely touched his brother before he was himself lying stunned upon the floor. As he arose, he drew a dirk; but Adam, who was cooler and stronger, caught his arm and wrenched the instrument of death from his hand. At the same time, he pulled the bell-rope, saying as he did so —

"Take my advice, and go; for, just so sure as you remain until the servant comes, I will send for an officer and have you arrested."

John stood glaring upon him with wicked, murderous eyes, evidently under the impulse to spring like a wild beast at his throat.

"Go!" Adam waved his hand. "From this hour you stand to me as a stranger."

"And an enemy!" John flung the words madly through his lips.

"Suit yourself in that; but go! I hear the servant's feet."

John stood hesitating for a moment, and then, with a long, wicked, devouring look at his brother, moved backwards to the door, and opening it passed out.

The servant came in immediately afterwards.

"Did you see the man who left here just now?" asked Adam.

"Yes sir;" replied the servant.

"Will you know him?"

"Yes sir."

"Very well. Now remember Henry, that if he calls again he is not to come to my room on any pretence. You understand me?"

"I do, sir."

"Say always that I am not in; and be sure to let me know when he calls."

"Yes sir."

"Should he attempt to force himself up here, restrain him, and if necessary hand him over to a policeman."

All of which the servant promised to do. But John did not call a second time. He had nothing to gain by a contest with Adam, and so stood far away from him, with hate instead of brotherly kindness in his heart.

His conduct towards his step-mother was of a character that soon gave her warrant to forbid him coming to the house, and she did not hesitate to accept the issue. And so, ere he had reached his twentieth year, the second son of Adam Guy, hopelessly enslaved to appetite and passion, and desperate in feeling, stood as completely alone in the world as if no kindred blood were in other veins.

To the husband of Lydia on her authority, had been paid the small legacy tendered in her father's will. Her proud spirit would have rejected this mean award; but the man who called himself her husband was of a different mould. He had married her for money, and now took whatever he could get; but took it in a spirit of angry disappointment.

The three older children effectually out of her way, and Edwin and Francis disposed of under the will, so far as a share of their father's property was concerned, Mrs. Guy, as residuary legatee to her husband's handsome estate, sat down in her calm dignity feeling that the long looked for time had come at last, when wealth and position were hers in actual right, and there was none to interpose word or act in contravention. Did no pity come into her heart? Did not the image of poor Lydia sometimes intrude itself and plead for a larger share in her father's estate? Mrs. Guy had no weaknesses. She was a woman of will and purpose; but not much ideality. If fancy did now and then conjure up what are called unreal things, they had no power to disturb the icy repose of her spirit. A fit counterpart was she for Adam Guy the elder; but, in meeting, she had proved the stronger spirit, and, by a transfusion of power, absorbed his freedom to a degree that made him almost passive in her hands. So, she accomplished her will, and in that accomplishment, set at naught all the cherished ends of a man, who felt that when he built his house on the solid foundations of gold, it was storm-proof and time-proof.

Oh, man! whoever thou art — wherever thou art — oh, man, in whose mind the thought of gold as the highest good shines ever as a star of brightest promise, take into your heart, and ponder it well, the life history we have given. Moral and mental causes work to corresponding effects, just as unerringly as manifest causes. Selfish ends defeat themselves by a law of compensation as inevitable as fate, for the germs of disaster are hidden in them at birth. In the degree that a man says in his heart, "Nothing but money!" just in that degree does he build on a false foundation; just in that degree does he put his gold in unsafe caskets. Avarice is blind in all directions but one, and there it sets watch and ward; but, while guarding approaches from this side with sleepless fidelity, enemies of whose existence no perception gives warning to the inner sense draw nearer and nearer, to work a final ruin, and they strike not until the thrust is surely fatal. If a man stood simply alone turning himself by a kind of spiritual alchemy into gold; or, by something of that process of displacement, and absorption of new elements, that we see in petrefactions, changing from a vitalized human spirit, into a dead form of avarice — the curse of a predominant evil love would rest alone with himself. When he went down in the sea of life, there would be few signs of shipwreck on wave or shore. But, not standing alone, and yet being false to nearly every scocial and home duty, how fearfully disastrous must the end be! Not in the devotion of our lives to any great leading purpose, do we secure final success and happiness; but in the devotion of our lives to an end that looks to other's good. Few, if any, so devote themselves, and few, if any, are successful and happy. — Still, the truth remains. Avarice and ambition are the most powerfully active of all selfish impulses, and drive some men onward in the world at almost the tempest's speed; but, avarice and ambition always leave ruin behind them — moral and spiritual ruin, over which the good mourn in unavailing sorrow.

THE END.